An American Library Association *Booklist* Editor's Choice

Praise for *Science Fare*:

"Perhaps you recall fondly your old chemistry set with its rows of chemicals ranged in a gleaming metal box; alas, 'There is nothing comparable on the market today.' This book is a contemporary resource that fulfills the same description."

—*Scientific American*

"Written with perception and enthusiasm, the text will encourage any parent to ask the right questions and find the best resources, from biology or paleontology to electronics and computers. An exciting volume for parents, librarians, and educators."

—*ALA Booklist*

"*Science Fare* is not so much a gift book as a gift of science to the children whose parents use it to engage their children in the fun of scientific inquiry and discovery."

—*Chicago Tribune*

"This is a treasure trove of materials and ideas to help parents engage their children in scientific discovery."

—*Library Journal*

". . . *the* book for every parent and teacher who wants to interest youngsters (preschool through middle school levels) in the exciting world of science."

—*Science Activities*

"This is an excellent book, full of ideas for parents to stimulate an interest in and to help introduce fascinating scientific ideas and theories to their children. . . . *Science Fare* is an outstanding science source book for teachers, parents, and libraries. . . . All in all, this book is an indispensable reference tool."

—*Appraisal*

"Because of its thorough coverage of subject and practical suggestions, *Science Fare* is an essential aid to parents, teachers, and librarians."

—*School Library Journal*

Science Fare

An Illustrated Guide and Catalog
of Toys, Books, and Activities for Kids

Wendy Saul
with Alan R. Newman

Introduction by Isaac Asimov

1817
HARPER & ROW, PUBLISHERS, New York
Cambridge, Philadelphia, San Francisco, London,
Mexico City, São Paulo, Singapore, Sydney

For Sylvan and Rosalie Saul
and
Matthew and Eliza Newman-Saul

 For information address Harper & Row, Publishers, Inc., 10 East 53rd Street, New York, N.Y. 10022. Published simultaneously in Canada by Fitzhenry & Whiteside Limited, Toronto.

Designer: Patricia Dunbar

Library of Congress Cataloging-in-Publication Data

Saul, Wendy.
Science fare.

Includes index.
1. Science—Study and teaching (Elementary)
2. Science—Study and teaching (Elementary)—
Information services. 3. Teaching—Aids and devices—
Catalogs. 4. Science—Study and teaching (Elementary)—
Bibliography. I. Newman, Alan R. II. Title.
LB1585.S28 1986 372.3′5044 85-45657
ISBN 0-06-181757-0 86 87 88 89 90 MPC 10 9 8 7 6 5 4 3 2 1
ISBN 0-06-091218-9 (pbk.) 88 89 90 MPC 10 9 8 7 6 5 4 3

Contents

Acknowledgments

This book would not have been possible without the help of many people. In some instances, a seemingly simple conversation contributed markedly to my understanding of the problems addressed here. I am grateful for the time and insights offered by scholars such as James Rutherford of the American Association for the Advancement of Science; Phillip Jones, mathematics professor at the University of Michigan; Anne Swanson, professor of chemistry at Edgewood College; John Jungck, professor of biology, Beloit College; and Jack Wilson, of the American Association of Physics Teachers. Help from the National Science Teachers Association has always been forthcoming, although Phyllis Marcuccio, Director of Publications there, has been particularly generous with her comments and time.

Each of the catalog chapters was critiqued by field specialists. Not only did these readers offer scientific expertise, but they were also able to convey a remarkable sensitivity to what makes a particular science appealing to children. Special thanks to David Bardack, paleo-biologist, University of Illinois Chicago Circle; Edward Burtt, ornithologist, Ohio Wesleyan University; Betsy Carlton, biochemist, Battelle-Columbus Laboratories; J. Edel, amateur radio operator, Delaware, Ohio; Dewey Moore, geologist, Knox College; Evan Hollander, electronics expert, Northrup Corporation Defense Systems; Wendell Patton, zoologist, Ohio Wesleyan University; Karen Richter, engineer,

Ohio State University; Andrea Smith, computer programmer, Nationwide Insurance Co.; Chet Snouffer, amateur weatherman and Olympic boomerang champ, Delaware, Ohio; Robin Taylor, educational software specialist, University of Maryland–Baltimore County; and Christina and Wolfgang Trautmann, physicists, Brookhaven National Laboratory.

Many children helped review toys, and we are grateful to them all. However, one young man in particular, Robert Amos of St. Anthony, Minnesota, deserves special thanks for his untiring, articulate, and thorough dedication to this project.

A number of professional librarians and teachers helped educate my taste as a reviewer of science literature. Rachel Alexander, former librarian at the Worthington (OH) Public Library; writer Vicki Cobb, Mamaroneck, NY; Lazar Goldberg, Hofstra University; Beverly Korbin, editor of The Korbin Letter, Palo Alto, CA; Pat Manning, Eastchester (NY) Public Library; and Bernice Richter, The Museum of Science and Industry, Chicago, IL, are represented in this book; it is a pleasure to be able to share their wisdom with others. The help of other children's literature experts is less obvious. Jo Osborne of the Worthington (OH) Public Library was an unending source of support and information. Iliana Ortiz, teacher in the Bronx, added an important dimension to these reviews, and David Gale, former student and literary friend, provided advice and criticism when it was needed most. Kathleen Weibel, the welcoming Director of Libraries at Ohio Wesleyan University, and her staff made academic searching a possibility.

Review copies of toys and books were generously provided by manufacturers and publishers, and, again, this book would not have been possible without their cooperation. The staff of the Toy Manufacturers of America and the staff of the Museum Store Association also provided invaluable aid.

On a more personal level, I wish to thank the teachers at Methesco Early Childhood Center, and Kristin Jones for the childcare that gave me the time to write. The organizational wizardry and encouragement of Reggie Kuhns worked wonders on the manuscript. And, finally, a thanks to my unnamed friends, whom I always imagined either smiling or shaking their heads after each paragraph.

Science and Children

by Isaac Asimov

There was a time when the study of the Latin language was considered an essential part of a liberal education. One of the reasons people advanced for this was that it made for good "mental discipline."

Latin has lost some of its ascendancy, and it is possible to go through high school and college today without taking any Latin at all.

Then what has happened to "mental discipline"?

Fortunately, we now have the study of science.

Don't think of science merely in terms of its content. It is possible to pile up the content indefinitely, and make that pile so difficult to grasp that only a gifted few can study long enough and intensively enough to master it all. Science, however, is so much more than that.

Science is, basically, a way of thinking.

It is careful observation. It is the study of the things about us in such a way as to see similarities and differences of many kinds: in structures, in motions, in behaviors.

It is collecting. It is gathering observations and noting them down, making a listing of things that seem to be so—of facts, in other words.

It is classifying—so arranging the facts that a certain sense can be extracted from them.

It is experimenting—asking questions of the universe. You deliberately set up an arrangement that will make something happen either this way or that way, thus creating a fact that you were not likely to see by mere observation, since the happening was not likely to take place of its own accord.

It is theorizing—seeing a connection among the facts, whether merely observed or obtained through experimenting. Once the connection is seen, all the facts may fall into place and become a kind of structure. It is like building a beautiful and sturdy wall of intricate design out of what had been a mere pile of bricks.

It is predicting. Once you have a theory—once you have built a connection among the facts—then you can see that certain events can't take place if that theory is correct. Other events, perhaps, *must* take place. You can experiment again and see if your predictions are correct. If they are, that strengthens the theory; if they are not, you've made a mistake and must start over.

Is this mental discipline? Obviously it is. You are taught to look at the world around you with disciplined senses, disciplined thought, and a disciplined mind. Your reward will be that you will find you live in a universe that makes sense.

Not everything necessarily will, of course. There will remain mysteries; there will probably always remain mysteries; but so many of these mysteries give way to careful thought that the universe seems to become friendlier and to be that much less a place of dark uncertainty and unpredictable whim.

And there is this difference between science and Latin: Both may represent mental discipline, and what's more, *science* is *fun*.

To solve puzzles is fun. To work with something which seems utterly confusing and to find yourself imposing order on that confusion, to see the darkness paling and becoming light, to be aware of knowledge supplanting confusion, is pleasure. You can be certain that anyone who experiences that moment of "Eureka!" is going to have an enormous impulse to jump up and click his heels, or to break into a dance of joy.

Nor is this the province of only a few gifted people blessed with unusual talent. It doesn't require years of preliminary study.

The method of thinking that is needed in science—the "scientific method"—is the same at all levels. Einstein used it in working out his theory of relativity, but a child can use it in studying the differences in the shapes of pebbles or in observing what happens when leaves fall from a tree. The subjects to which the scientific method is applied are infinitely different, but the system of thought is the same.

In fact, it is to children that the scientific method *should* be taught, for it must be instilled early. If a child grows up without this mental discipline and becomes an adult without having learned how to think in a systematic way, it may be too late to begin then. Without a proper introduction to disciplined thought in the malleable years of childhood, a person may well become used to haphazard conclusions, to fallacious judgments, to blind acceptance of superstition, and never able to break away.

But isn't it possible to teach children to think properly without recourse to science?

Literature, history, art, philosophy, even Latin—there are so many areas of learning and study that require careful and disciplined thought. Why do we have to pay particular attention to science?

The answer is that science, more than any other field of study, is of peculiar importance to all of us—to every person on earth, young and old, male and female, city-dweller and country-dweller, learned and ignorant.

It is science that has built our present-day industrial civilization and that has created all the problems that come with such a civilization—overcrowding, pollution, noise, danger of nuclear war, and so on. It has also created all the comforts and securities we now have that our great-grandfathers did not have—rapid, comfortable transportation; instant communication; advanced medical treatment; new forms of entertainment; and so on.

Science, wisely used, can continue to give us more comfort, better health, longer life. It can solve the problems that face us.

Science, unwisely used, can make the problems worse and can give us newer and even more horrible ways of destroying ourselves.

The world must be able to choose between the wise and unwise use of science; it is literally a matter of life and death for all of us. And just as war is too important to leave to the generals, and government is too important to leave to the politicians, science is too important to leave to the scientists.

There must be an informed public opinion that presses for wisdom and stands against folly. But how can we choose between wisdom and folly? How can we be sure which is which?

There is, we must admit, no certain way. There is always the possibility of choosing a course that seems sensible and rational, yet turns out to produce unlooked-for dangers.

This, however, we can be sure of: If the people of the world know nothing of science, they will never be able to make a wise decision among alternative policies. With our eyes open, it may be difficult to choose the right path, but with our eyes closed, it will be impossible to do so.

Mind you, this is not to imply that everyone must be a scientist. Not everyone has the talent for it; not everyone would want to be; and the world has no use for four and a half billion scientists. We need shoemakers, too, and pianists. Still, you don't have to *be* a scientist to know enough about science and the scientific way of thinking to participate intelligently.

There are few people sufficiently equipped by nature and sufficiently willing to endure long hours of practice to become professional baseball, football, or basketball players. There are, however, tens of millions of Americans who are sufficiently interested in these games, and who are willing to apply themselves in order to learn enough about the rules and techniques, to be intelligent spectators. They are thus enabled to enjoy the games and to have reasonable opinions as to what ought to be done at various crisis points.

In the same way, there should be vast numbers to people who will learn enough about the rules and techniques of science to enjoy the greatest game of all—that of the human mind pitted against the intricate mysteries of the universe—and to have intelligent opinions as to what ought to be done at various crisis points in that game.

And to have a world populated by such people means that we must start with the children.

Furthermore, in years to come we are very likely to need more scientists than we now have. I am not referring only to the great researchers who win Nobel Prizes and other high honors—the field-marshals of the scientific army. These are usually drawn from people who are in love with science from youth and who need no encouragement beyond that which their own temperament supplies.

Behind them, however, are large numbers of researchers who do worthy work but go without newspaper-headline rewards. And behind those are the unsung people who keep laboratories and hospitals going, who run tests and make decisions every minute of the working day—the sergeants and privates of the scientific army, who are less highly motivated, perhaps, and who need more encouragement to find their place in the ranks.

Count them all in and they are a huge host. It is frequently said that 90 percent of all the people who have worked in the various fields of science in one way or another are *alive today.*

If all goes reasonably well, the future is going to see us develop science and its applications faster than ever. Computers and robots will fill the world and they, along with people, will move out into space—to the Moon, to Mars and beyond. We will need more and more scientists, technicians, engineers, to add to the army that will be required to run the far more complex world of the future. Where are they to come from?

It is not likely that society can simply count off a million people at random and say, "From now on, you must work in science." If they didn't have the proper introduction to science while young, if they haven't learned to think like scientists, the best will in the world won't make them scientists.

Let me emphasize this. We can only obtain the scientists we may desperately need in the future from people who, from childhood, have been interested in science and have learned to think like scientists. This, then, is another major reason why we ought to make a great effort to introduce science and the scientific method to children.

But how is that to be done?

Science can be introduced to children well or poorly. If poorly, children can be turned away from science; they can develop a lifelong antipathy; they will be in a far worse condition than if they had never been introduced to science at all.

How useful it is, then, to have a book such as this one, which discusses methods by which science can be made accessible to children. It takes up every aspect of the science learning experience that has, or can be made to have, an entertaining aspect. It tells you how to go on outings, about the hands-on museums and summer camps that are available, where to obtain the most useful toys and games. It deals with something as simple and seemingly artless as the best way of answering questions, and with something as formidable as science fair projects (and how to survive them). In the second and final section of the book, you will find a catalog that will tell you what books and activities are suitable for different ages—and even whether they are suitable at all.

Let me conclude by saying that children are individuals, and that what works for one may not work for another. It is important, then, to consider the particular child and to note his or her particular bent. In my own case, for instance, I became interested in science through my more or less accidental encounter with science fiction magazines when I was nine years old.

Science Fare

All names that appear in SMALL CAPS constitute proprietary brand names or trade names.

1 The Problem and the Plan

SOUND FAMILIAR?

Last year the Lawsons purchased a telescope for their eleven-year-old son, Tim. It was his "main" Christmas present, an item he'd anticipated for months. Christmas day was spent boning up on the constellations and awaiting the glory of the winter sky. Unfortunately, the lights of the neighborhood made interesting observing impossible.

A month later we visited the Lawsons on their Wisconsin farm. High atop the ridge, with no neighbor in sight, we tried the Christmas gift again. The kids were excited. We mounted the telescope on the tripod, and Tim peered through the eyepiece.

"It's all blurry," he complained.

"You need to focus," replied his father.

"I *am* focusing and it's *still* blurry." We could hear the frustration in his voice.

Tim's father tried next. In fact, we all tried. After forty minutes of unsuccessful fidgeting in the bitter cold, we came indoors. My husband, an amateur astronomer and a man very reluctant to disappoint an eleven-year-old, finally came forward with an explanation. Apparently the telescope had come with a lightweight tripod, which made the system extremely sensitive to ground and wind vibrations. The stars appeared more blurred through this mechanism than they did to the naked eye.

These events left several people feeling angry and inadequate. Tim was angry because his parents had bought him a "cheap" telescope. (This was untrue, in fact—it was an expensive gift for an eleven-year-old.) Not being a terribly resilient type, he shuffled off to bed mumbling something distressing about his future.

My husband was upset too. There he was, with three educated, otherwise sensible adults who, had he not been around, might still be fiddling with an inadequate item. Our tendency, he realized, was to blame ourselves rather than the mechanism. Furthermore, because we did not understand the limitations of the equipment, we were ready to retreat from scientific play.

Tim's parents were equally distraught. They had investigated telescopes as well as could be expected and had chosen the one in their price range that had the greatest degree of magnification. But they didn't know the right questions to ask. And no guide to science toys and equipment was available.

* * *

Marta saved her allowance for 24 weeks in order to purchase an electrostatic generator. Finally her gadget arrived, and it was wonderful. She could turn a wheel which pushed two gizmos with brushes in opposite directions. The static electricity these brushes generated was stored in a condenser—enough static electricity and, whamo!, sparks were discharged.

Marta watched with delight as her machine sparked. Then she placed the disc attachment on the shaft and watched as it whirred. Then she generated more sparks. Then she called in her parents to watch the sparks. (They, too, were impressed.) She called in her brother to watch the sparks. She called in her friends to watch the sparks. And at last, after sparking and whirring for a good hour and a half, she put the machine away . . . forever.

Marta's parents wanted to help their daughter extend her basic interest in physics, but how? What experiments or what books might they have suggested? What conversations or questions might have inspired further exploration and excitement? They didn't know where to begin.

SCIENCE EDUCATION: WHOSE BAILIWICK?

Science and technology are important to us all. We drive cars, we look to the medical community for cures for our illnesses, we work with word processors and calculators. And yet our society seems divided between scientific and technological experts and the rest of us who naively use their products and processes, hoping for the best and praying for safety.

In the past few years we as a nation have come to recognize that this is a problem: we have even given it a name—"the crisis in science and mathematics education." And because solutions to educational problems are generally seen as school matters, we look to our teachers, principals, and boards of education for solutions. School personnel, in turn, direct attention to curriculum reform and better teacher training.

Please do not misunderstand: curricula and teacher training can and should be upgraded. But parents concerned with their own competence and that of their children may be considering the problem of science education too narrowly. Informal scientific and technological education can be as important as formal studies. For instance, researchers surveying former Westinghouse Science Talent Search winners found that students first became interested in science and math in grade school—or earlier. And most claimed that their families, not their teachers, provided the early help and excitement that led them to pursue scientific studies.

It is a mistake to assign all things intellectual to the schools. In so doing parents deny themselves and their children a variety of pleasures. Children who share ideas and learning experiences with their families grow to see honest, uncompulsive work as satisfying, the kind of experience against which other good things are judged. Working together on a wild-flower collection, they discover, is every bit as much fun as collecting stickers or playing a board game.

Children involved with informal science also benefit from the regularity of the play—often a child understands something amazingly different the tenth time he or she performs a particular experiment than was obvious the first time. Being able to manipulate objects in a new, more competent way is also an important gratification.

In addition, informal science affords children an all-too-rare opportunity for unhurried and fanciful exploration. Jennifer can study a single petal of a tulip without having to pass it on to the next kid in the row, without being interrupted by recess, without worrying about what grade she will get on her tulip composition.

And finally, at-home study allows parents to give to science the kind of attention and caution they believe the subject deserves. Through concrete experiences, thoughtful adults teach their young that learning is a lifelong process. Children who grow up in such an environment come to understand their human past and present, and its complicated and fascinating relationship to nature.

ENCOUNTERING SCIENCE: HOW THIS BOOK HELPS

There are four general ways that children encounter science in our society. The most obvious and most frequently overlooked stimulus is completely natural. Living in a world where objects fall when they are dropped, where the sun rises and sets, where people invent machines and talk about fixing others, children naturally ask questions and order information in naive but scientifically interesting ways. There are a surprising number of books for adults, many of which are listed in this volume, that suggest activities to help children attend to the natural world. But the activity approach is akin to following a script, rather than engaging in a conversation. This book is designed to help you become better attuned, and better able to respond, to a child's natural search for scientific meaning.

The second way that children learn about science is through institutions designed to teach science. Schools and museums are the best-established of these forums. Chapter 6 discusses school science fairs and offers suggestions on ways parents can become more meaningfully involved in their children's school science program. Community resources—museums, zoos, aquaria, and planetaria—are discussed in Chapter 5. These are places in which people have given considerable thought to what is displayed and how children learn. Moreover, too little has been said about the art of being an active viewer.

The third way children learn about science is by playing and working with toys and equipment designed to encourage or capitalize on their scientific and technological interests. There are many excellent products available, but few can be found in your local toy supermarket. I have searched and scrambled to provide an extensive, selective list of such items and where they can be found. Moreover, comments from children, scientists, and educators who have played with these items are included, along with detailed information on what works, what doesn't, and what you can add to make an acceptable item good or even excellent. Each review is written to answer the question, "Who might find this particular product appealing?"

Frequently scientists refer with fondness to their early attempts at making rockets or taking apart a discarded radio. Tinkering often leads to good questions, data collection and interpretation, knowledge seeking, and other attributes integrally related to science. Furthermore, science paraphernalia provide opportunities too frequently overlooked in the classroom. A child who has difficulty sitting still for twenty minutes in a social studies class may spend hours at a time mucking about with an electronics experiment. In short, science "things" have received too little attention.

This analysis is supported by research. In a recent report undertaken by the Federation of Behavioral, Psychological, and Cognitive Sciences for the National Science Board Commission on Pre-College Education in Mathematics, Science and Technology, investigators found, not surprisingly, that interest in science was important in determining science-related success. They speculated that in-school interest may well depend on prior knowledge and experience. Girls, for example, who in general have had far less informal experience with science toys and hobbies, could be at a distinct disadvantage in the science classroom.

There is, however, a problem with most science toys and sets: they do little to provoke or answer questions born out of the hands-on experimentation and play they provide. This need for explanation leads to the fourth area I identify as important in learning about and appreciating the physical world: non-fiction. In this sense, libraries and bookstores become important agencies for promoting more and better science. Unfortunately, book collections which appear outstanding in other respects are often sadly lacking in information of scientific or technological interest. Chapter 7 includes many suggestions for evaluating science books as well as bibliographic references for keeping up with what's excellent or new in science literature. Furthermore, every chapter includes specific recommended titles. Fortunately good informational literature is usually pleasurable reading.

Finally, detailed information about finding answers to questions of scientific interest is provided. For those people who are not schooled in science, or for those scientists not altogether familiar with the reference tools designed for children, Chapters 4 and 7 should prove particularly helpful. The ability to gather information and locate clear explanations is key to scientific appreciation.

In sum, this is a book for adults ready and anxious to attend to children's natural interests, lobby for more and better science in educational institutions, and provide young people with the toys, equipment, and books they might use to better understand, explain, and question the world in which we live.

2 Approaches to Science Education, or, Methods Can Make a Difference

Because science education has generally been viewed as a school matter, descriptions of how science should be taught have focused on textbooks and curriculum packages. But consider the variety of approaches to science that might be taken in an informal setting.

GEE-WHIZ SCIENCE

The flashiest, most titillating approach, and the one most frequently criticized by thoughtful educators, can be called "gee-whiz" science. Imagine, for instance, a book of world records filled exclusively with science facts—isolated bits of information about extremes, taken out of context and designed to shake the reader. I understand why people in science disavow the gee-whiz package; it doesn't reflect the importance of theory, an appreciation for how things come to be known, or a sense of the interrelatedness of events and conditions.

On the other hand, even those children who take science seriously, the science fair crowd, love the gee-whiz stuff. The difference is that they see themselves as contributors to the gee-whiz of tomorrow, and are better able to appreciate the gee-whiz of today. As one young man put it, "I spend hours out there with my telescope, thinking of all that I can't see. That's why I like to look at double stars. It's proof that [gee-whiz] there really is a lot more—lots, lots more."

Gee-whiz science is best seen as an entry point, not an end in itself. Non-scientists need to realize that the gee-whiz isn't like icing on a barely palatable cake, but rather one amazing moment that may lead, with some work, to other, often more amazing moments, and to decisions that can affect their lives and the lives of their children. At its best, gee-whiz is the visceral appreciation of fact and order.

Frank Press, president of the National Academy of Sciences, is one expert who doesn't want "astronomy without stars . . . botany without the flowers . . . geology without the mountains and valleys." Science education, he claims, is not true to human nature when it separates that which people value from that which they seek to understand. As adults concerned with children and science, we need to remain vigilant: unless science is perceived as being passionate as well as logical, as part of a larger attempt to describe and grapple with existence, we have failed.

The sun was setting gloriously over the Connecticut hills when an astronomer friend looked up and remarked simply, "What a marvelous planet." I was startled to see the sky as I had never seen it before—realizing that the blue might have been yellow, grateful for the sun's warmth which in some other world would be absent . . . gee whiz.

LEARN-THE-FACTS SCIENCE

Another approach to science can be loosely termed "learn-the-facts" science. At its worst learn-the-facts science leaves children's heads swimming with meaningless and unconnected bits of information and formulae, knowing, for instance, that there are clouds called cumulus, cirrus, stratuscumulus, and nimbostratus, but unable to remember which are which, or even why one bothers to identify clouds.

As in most things, there are both good and bad reasons for the continuing emphasis on science as a body of information to be recalled. Adults who are shaky on a real understanding of a subject can steady themselves with these isolated demonstrations of memorized might. In other words, a scientific vocabulary can be used as a form of control.

In addition, factual information has great "brag value." While virtually all eight-year-olds can recite the alphabet, few are able to rattle off the noble gases from the periodic table. Although such a recitation may be a great party stopper, in fact it indicates nothing about the child's ability to observe or reason.

Even though the memorization of facts can be misused and abused, do not completely disavow the idea of collecting and storing information. Facts are important—they give form and precision to that which we understand. If, for instance, you know that those pointed toenails on a bird are called talons, you can look up "talons" in a dictionary or encyclopedia and find that they are used in ways that toenails are not. You may also realize that animals other than birds have talons, which may lead you to wonder about the connection between our toenails and talons. Did toenails ever have a purpose? Do they still have a purpose? Are they really made of the exact same material as talons?

Similarly, information makes certain investigations possible. Facts enable people to define a problem and share information with others interested in similar matters. A junior astronomer needs to know how to locate key stars, planets, and constellations in order to read a star map and enjoy a telescope. A youngster working with an electronics set must understand that "negative" and "positive" do not mean bad and good in this situation.

Facts are also intrinsically related to questions. There are few things more frustrating than generating lots of good questions, being congratulated on developing good questions, and then having no resources for answering them, at least in part. As a child, Sharon was bothered by the idea of colors: "I knew that two people could look at a building and say that it was white, but I was bugged by the possibility that what you saw as white might be different than what I see as white. When I learned about the spectrum and the functioning of the eye, it was a real relief."

Furthermore, facts are closely related to language, and language is its own pleasure, its own power. I like the fact that the end of the bannister is called a newel post, not because I want to learn anything more about newel posts, but because it's a shapely (as opposed to a clumsy) expression. Precise language also helps a young person share his/her findings with more advanced investigators; thus, a vocabulary becomes a form of initiation into a scientific community. Facts are also the basis for any taxonomy. The four-year-old child who derives great satisfaction from being able to distinguish between a Volkswagen and a Cadillac is equally pleased to know that a robin and a cardinal are different.

In addition, having facts in storage is convenient, significantly convenient. For instance, if Alec is trying to map out the bug activity in his backyard and has to look up each bug he sees, as well as each plant on which he sees each bug, the activity becomes cumbersome and unreasonably difficult. If, on the other hand, he knows that there are two bugs on the goldenrod, and that one is a monarch butterfly and one a tiger beetle, he is in a much better position to follow what happens in the next moment when the butterfly heads for the mullein and the tiger beetle continues to hunt out (what could it be?) on the (what *is* that fat podlike thing?).

Finally, facts can help contextualize seemingly dry, quantitative information and make it interesting. Einstein's theory of relativity is brilliant not only because it tells us something important about our universe, but also because it was conceived at a time when people read by gas lamps and automobiles were seen as extraordinary. Our student friends don't really object to facts, but rather to boring, displaced, irrelevant facts. The best teachers, both formal and informal, help to make facts interesting, relevant, and provocative.

THE THEORETICAL APPROACH

The third approach to teaching science can be termed "theoretical." In this case, it is the job of the adult/expert to explain significant ideas, while the child/novice is expected to ask questions and find examples which illustrate the grand theories presented. Most non-scientists do not even consider the theoretical approach as an option. How does one explain something like magnetism to a child without having a better grasp of the vague but relevant details most adults remember (something about a molten core in the earth, and the attraction of metal by an undefined force located in polar regions). Recent research, however, strongly indicates that children, *all* children, intent on sorting out events and putting information into categories, develop their own, often incorrect, theories or ideas about how the universe is put together. Often these misconceptions are terribly difficult to dislodge.

In short, there is a problem here. Adults without a background in science typically avoid offering theoretical explanations to children, but children, in need of hierarchical ways to order information, invent explanations which are frequently wrong. I have several suggestions for dealing with this difficulty. They are all based on the notion that what you say about science is as important as what you do.

- Recognize places where cogent explanations to events can be found—museums, libraries, magazines, and so forth.
- If you are uncomfortable offering explanations, introduce your children to others who are talented in handling these kinds of conversations. For instance, Jenny's father commented to an "explainer" at a museum, "Jenny read that molecules don't have any color, but when you get a lot of them together, like in this orange carpet, they do have color. Can you tell us why?" In this case, Jenny's father is modeling information gathering, and Jenny is able to see that her innocent question was worthy of her dad's attention.
- Show your children that you value theories that are "intelligible, plausible, and fruitful." Test your own ideas (make sure to throw in a few mistaken ones) using these criteria. And see Chapter 4 for more on this subject.
- Recognize that children learn from watching the behavior of others. If you are interested in revising your own theories, and if you enjoy learn-

ing new things, your children are more likely to be open-minded and creatively skeptical.

Theory should be part of the science education of all children. Providing explanations only for those deemed talented surely runs counter to our democratic traditions. Cookbook science does not enable children to give creatively to, and take thoughtfully from, a scientifically and technologically sophisticated world. As one physics teacher put it, "All those guys who know how to take their engines apart and put them back together deserve to know how and why their cars work."

HANDS-ON SCIENCE

The fourth method used to teach science might be called the "hands-on" approach. While the gee-whiz family is calculating how much Harold Junior would weigh on each of the planets; while the learn-the-facts family is engaged in a game called "Name the Elements"; while Mr. and Mrs. Theory are explaining to little Dorothy about the second law of thermodynamics; Mama Hands-On is setting up an experiment for Betty and Rob. In this instance the children are given a basin filled with water, a roll of aluminum foil, and a lot of assorted junk. The children are challenged to design a vessel out of aluminum foil to hold the junk. Experimenting with possible shapes, they learn something important about floating as a principle, and about ship design as an example. They also learn that the junk they put on their foil boats has to be properly balanced.

I like hands-on science for several reasons. First, there is a hands-on, messy, imperfect aspect to adult science; to teach the subject without acknowledging this element would be inaccurate. Second, in order to invent and discover, children need materials to invent and discover with. Moreover, hands-on, discovery, and process-oriented activities call for creativity and logic, as well as open-mindedness and care—important attributes all. And last but not least, the hands-on approach takes seriously the idea that science is something everyone can do and enjoy.

Despite these benefits of hands-on programs and activities, there are two problems with the approach. The first is that hands-on science is more difficult to organize. I remember visiting a school in which the tadpoles for a project already underway arrived dead. Furthermore, there are logistical problems in keeping sufficient jars, aluminum plates, metal scraps, and measuring devices ready and available. For anyone used to total control, these difficulties may seem insurmountable. If, however, you are interested in promoting science, you need to realize that these difficulties are what real science is all about.

And second, hands-on science does not necessarily counter naive misconceptions. Deborah imagines that her plants have taste buds. She wants to do an experiment to find out if they like lemonade better than water. If the geranium in question flourishes on the combination of sugar and acid in the lemonade, Deborah may well feel that her theory was borne out through a real experiment. Well-prepared parents and teachers learn to listen carefully to the ways in which children explain what they see and do. An ideal teacher or parent would be scientifically literate enough to point out misconceptions and steer children to appropriate sources for further study.

In sum, hands-on science is not a total substitute for reading or learning from others. Hands-on science *is* good because it can provoke questions, or focus interests, or demonstrate certain principles. It is also fun. It should not, however, replace informed conversation and reading.

All Hands-On Science Is Not Equal

Lots goes on in the name of science which, in fact, leads children away from a better understanding of the physical world. Here's an example.

Two mothers respond to their children's interest in clouds. The first buys poster paper, an assortment of drawing implements, cotton balls, and glue. "All right, kids", she begins. "What do clouds look like?"

With little effort she elicits the desired response: "They are fluffy and white."

"And what did Mommy buy that's fluffy and white?"

"Cotton balls," the children chant back in unison.

Ways to Tell a School You Value Science Education

A parent, especially a parent who is uncomfortable with science, cannot walk in off the street and hope to redesign a school's science curriculum. There are, however, positive, nonthreatening ways to tell teachers and administrators: "I value science, I think that it is important to learn, and I'd like to support your efforts to teach it."

- At a parent-teacher conference, ask how your child is doing in science. What does the teacher think of his/her problem-solving abilities? Is she/he able to manipulate variables? How does he/she do in lab work? Take it from a former teacher—if teachers know that parents are concerned about a particular academic area, they are more likely to develop related lessons.
- If you are a member of the school's parents organization, ask that some of the money you raise be earmarked for a science-related purchase or to pay for in-service science education. Think specifically of your school's needs—a computer that children use one at a time may not have as much impact as microscopes for an entire class.
- If you are involved in a school book fair, make sure that science-related books are available and attractively displayed. Set up a book on microscopes next to a working microscope, or a book on gerbils next to the class gerbil.
- If your child has described a lesson you think is particularly good, tell the teacher, and ask if there is anything you can do to help with more activities of that sort.
- Encourage your child to bring a science-related "show and tell" to school.
- If the school invites volunteer work, offer to do a science-related bulletin board, set up a science display in the library, or arrange for a speakers program.
- If you are asked to attend field trips, say that you are particularly interested in science/technology trips.
- Ask that a teacher willing to set up a science club or interest group be compensated as the coach of an athletic team is compensated.
- For class gifts, you can't beat a subscription to one of the following magazines:

Audubon Adventures, bimonthly, $20 for members of the National Audubon Society (National Audubon Society, 613 Riverside Rd., Greenwich, CT 06830).

This classroom publication features a nature newspaper for kids and an excellent "Leader's Guide" for teachers. Both are well informed and well written.

Exploratorium Magazine, 4/year, $10 (3601 Lyon St., San Francisco, CA 94123).
This 30-page thematic publication explains the science behind everyday activities in ways that are hard to resist. Their award-winning issue on mapping or their recommended "bubbles" activities are models worth emulating.

Kind News, 5/year, $10 (The Humane Society of the U.S., 2100 L. St., N.W., Washington, DC 20037).
Articles, puzzles, projects, cartoons, and games are found in this 4-page, class-oriented publication. Available in editions for grades 1–2 or 3–6, 35 copies per class set.

Naturescope, 5/year, $18 (National Wildlife Federation, 1912 16th St., N.W., Washington, DC 20036).
Nature-related lessons, planning tips, craft projects, stories, plays, mini-courses, worksheets, and more.

Science Activities, 4/year, $35 (4000 Albemarle St., N.W., Washington, DC 20016).
This useful journal, chock full of hands-on projects, book reviews, and classroom-oriented ideas, uses a thematic format. Back issues can be purchased separately for $5 apiece.

Science and Children, 8/year, $32 (for members of the National Science Teachers Association, 1742 Connecticut Ave., N.W., Washington, DC 20009).
Directed to an audience of elementary-school teachers, this attractive journal includes classroom ideas, book and media reviews, and discussions of national issues in science education.

Science Scope, $5 (with a subscription to any other National Science Teachers Association journal).
A lovely bonus gift for a junior high school teacher who already subscribes to either *Science and Children* or *Science Teacher.*

Add to this "wish list" other excellent publications listed in Chapter 7, such as *Appraisal and Science Books and Films,* or any of the children's science magazines with teachers' editions, such as *Ranger Rick,* found in specific catalog chapters.

This mother encourages her offspring to draw an outdoor scene, and praises them as they glue clouds in appropriate places in the sky.

Our second mom has a much better understanding of science. She takes out a tea kettle and a cookie sheet, heats the water in the kettle on the stove, and literally demonstrates how clouds and rain are created. As the steam changes to water vapor and collects on the cookie sheet, the children see water droplets form. They beg for a repeat of the demonstration, and chat excitedly about the way rain clouds differ from other clouds and what happens when a cloud settles on the earth.

The point here is not that demonstrations work better than drawings, but that the first mom's caricature of a cloud is, in effect, anti-scientific. Her children were never invited to look out the window, or to study photographs of clouds, or to find meaning and order in their environment. Clouds for this parent were merely decorations in the sky, and she blithely passed this attitude on to her children.

AN ECLECTIC MODEL

In the best of circumstances, the four approaches weave in and out of any scientific encounter. Kite flying, for instance, can be a glorious gee-whiz experience; it's amazing to feel the tug of the string, and to consider that you are holding on to that diamond in the sky. You can also learn facts through kite flying. Because you wish to take your kite to the beach, you listen in a new way to coastal information about wind velocity. While flying the kite, questions naturally arise: "Why do some models soar higher than others? Under what circumstances?" Suddenly you have an idea: "Let's see what would happen if we made the tail shorter." In moments a hands-on experiment is under way. Later, an excellent book from the library might suggest design modifications for the next kite flight and provide answers to the questions raised that afternoon. This event could be called a perfect scientific encounter, or it could be called making the usual "get the kite up, take it down" routine more interesting and more fun.

QUESTION: If I take my child kite flying, ask the right questions, and follow up the experience with a trip to the library, is she more likely to get into med school?

ANSWER: Statistically you may have improved her chances, but be assured, for better or worse, you're dealing with someone real who does not respond in exactly the same way another real child does to the same events. Learning science isn't like advancing on a conveyer belt and being stamped with formulaic imprints. Please use this book as a reference work; it's up to you to choose those items and ideas that make particular sense in terms of the children you know and love.

Furthermore, don't push your young ones. There's no rush to get your child to point X by a certain time. Apparently Einstein believed that his scientific talent was largely borne out of his "slowness"—because he began to think about the physical world at an age when other kids took it for granted, he didn't accept natural forces ordinary and commonsensical. Even if all children were exposed to the same scientific encounters at the same age, they would not all end up thinking the same way . . . thank goodness!

If children learning science are not like bits of raw material being turned into uniform finished products, what are they like? Learning science may be better compared to taking a journey. People naturally plan for trips in different ways and respond differently to their surroundings once they arrive. Some people read every travel brochure available and have every minute of their stay planned in advance. Others simply take off, ready to be surprised by what they find. No one expects us all to agree on what journeys we find most enjoyable, or what places are best to visit at a particular stage of our lives.

If learning science is like a journey, then we—parents and teachers, librarians and youth group leaders—can be seen as guides. As such, we point out items of interest and answer questions as they arise. On the other hand, we don't expect that all children on this journey will respond positively to the same events, circumstances, and stories. As guides, we also hope to learn from our visitors—they each have experiences and perceptions that are exciting and unique.

Moreover, as guides we realize that no one person is able to see and understand everything in the rich environment we provide. What makes our job interesting, in fact, is that we are continually seeing the same things in different ways. There are trade-offs; if one is looking at the ceiling of a building, one cannot, at the same time, study the walls and floor. Finally, as guides we must realize that we don't know everything about a particular excursion; our hope is that we have interested our visitors enough so that they will read more about their experience, talk with other visitors, and return again to this spot, perhaps with a different guide next time.

Science Opportunities for All

There are many reasons why we as a nation are not scientifically literate: the presentation of science is often inadequate, teachers themselves aren't well enough schooled in science to produce good programs for the children in their charge, parents and other informal educators are not doing enough with science. But one other major problem looms ominously on the horizon. To date, science and technology have generally been considered subjects for able-bodied white males. Girls and women, Blacks, Hispanics, Native Americans, and the handicapped too often see themselves as outside science. This is a tremendous waste.

I wish to be specific rather than rhetorical here. Advocates for these underrepresented groups can do some or all of the following:

- Be particularly conscious of inequities in the home. Are you buying science toys and books for *all* your children? Are certain members of the family getting more science and mathematics attention than others? Are your expectations for your children appropriately ambitious? Do your children feel welcome and encouraged to participate in school science experiments and events? Research by Jane Butler Kahle of Purdue University, and others, indicates that even in the same courses, females and Black males have less experience actually performing experiments and taking field trips than white males.
- In groups which specifically cater to girls, minority children, or the handicapped, see that science and technology-oriented projects are introduced. These projects should be designed to evidence mastery. Once children see that they can succeed in a non-threatening setting, they are more likely to participate in school projects.
- Look for appropriate role models.

  Girls need to see that there are successful women engineers, physicists, car mechanics, and so on. Biographies and personal contacts have both been important to aspiring female scientists. An organization called "Women in Science," with local chapters nationwide, has been very generous in providing a forum where working women scientists and girls interested in science can meet. The American Association for the Advancement of Science (1176 Massachusetts Ave., N.W., Washington, DC 20026) also has an Office of Opportunities in Science, which serves as an excellent resource for girls and women interested in scientific studies.

The handicapped can contact the A.A.A.S. Project on the Handicapped in Science, a powerful resource center disseminating science education and employment information. (Write c/o Dr. Martha Ross Redden, Director, Office of Opportunities in Science, 1776 Massachusetts Ave., N.W., Washington, DC 20026). This group also publishes a "Resource Directory of Handicapped Scientists" who are willing to be contacted for help and information. Both attitudinal and technological solutions to individual problems are shared through this impressive network. A compelling book to share with disabled children, *Able Scientists, Disabled Persons* by Phyllis Stearner, is available through the Foundation for Science and the Handicapped, Inc., 154 Juliet Ct., Clarendon Hills, IL 60514. The National Science Teachers Association (1742 Connecticut Ave., N.W., Washington, DC 20009) also has a Special Education Committee. Remember, handicapped children have a legal and moral right to science instruction, laboratory experiences, and field studies.

Minority students are also served by the A.A.A.S. Opportunities in Science group. Their information, helpful to both individual children and teachers, does not lump together all minority children, but rather discusses cultural differences and provides information on specific, relevant programs. Information on scholarship opportunities and summer research programs is also located here.

- Listen carefully to your children as they describe their experiences with science. Elizabeth Fennema, a professor of mathematics education at the University of Wisconsin, notes that girls tend to attribute success in science and mathematics to luck or good teaching, whereas boys attribute their achievements to ability and hard work. Generally, people do not pursue a subject if they feel they were "just lucky" to get a good grade. Competence and confidence go hand in hand.

- Make sure your standards are the same for boys and girls. Research indicates that teachers and parents have lowered expectations for girls and minority students. "Easy grading" or false compliments do not serve anyone's interests.
- Look for examples—word problems, machinery, experiments—with which female, minority, and handicapped young people are comfortable. A sewing machine, vacuum cleaner, or wheelchair, for example, can be used to teach various physical principles. Instead of taking sodium chloride out of a mysterious, professional-looking bottle for a given experiment, try adding salt from a recognizable container.
- Discuss the importance of scientific contributions to the lives of individuals and communities. Many people in these special groups are interested in making a social contribution, but they see science as being "thing" rather than "people" oriented.
- Encourage spunk when you find it. Most of the scientifically successful female, minority, and handicapped young people to whom I have talked took up science as a kind of challenge. One young woman raised on a farm said that when her father slaughtered animals, he'd hold up the guts to frighten her. To show her resistance, she would calmly take these entrails into the kitchen, dissect them, and lay them out carefully on the kitchen counter. Then she began going to the library, first to identify what she had found, and later to understand about the organs' functions.
- Last but not least, remember that minority, handicapped, and female children have a variety of social conventions working against them. To combat prejudice, parents may have to become aggressive advocates. Take your children on science-related trips, provide lots of hands-on experiences, and show enthusiasm and support for their efforts to achieve and understand.

3 Encouraging Inquiry and Understanding

DEVELOPING INQUIRY SKILLS

In science, certain skills and analyses are valued more than others. Children who, even on an elementary level, are able to observe, classify, predict, estimate, measure, manipulate variables, and explain are at a tremendous advantage in approaching scientific and technological activities. An interest in objectivity and a skeptical attitude also serve young people well. Experience with, and an appreciation of, these skills can easily be learned at home.

Observing and Classifying

Betsy, age seven, called to her mother to come and see the piles of dandelions, pebbles, pine needles, twigs, and leaves she had sorted in the backyard.

"Great, Betsy," said her mom, and returned to the laundry.

Ten minutes later the child's call was heard again. This time she had divided the stuff into two piles, one of "things from trees" and the other, "things from the ground." Mom was becoming more interested, and tried a sorting scheme herself. Betsy guessed that one pile contained living and the other non-living things.

Rather than save the piles of wilted dandelions and driveway stones, Mom told her daughter that she would show her how scientists write down similar

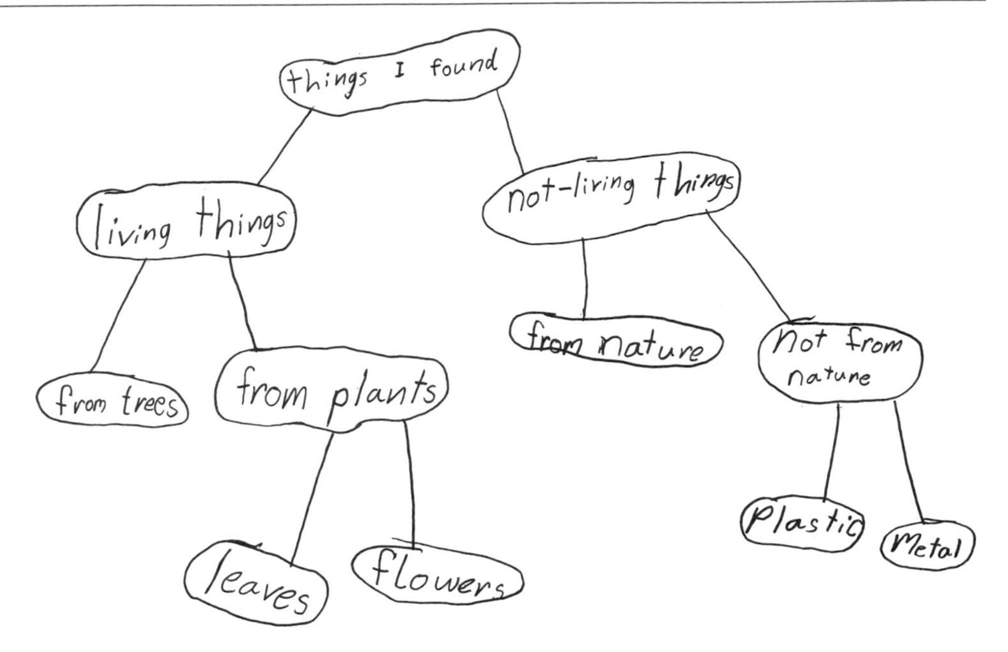

information. For the next several weeks schematic drawings of this type were compiled.

Similar everyday activities can become occasions for scientific conversations and experiments. While traveling through the supermarket, let your children locate items on your shopping list, and discuss why stores organize foods as they do. "Why are certain items refrigerated?" "Which things that we bought today grow on trees?" As you stop to read labels, explain what you are looking for. Tell children that bruised fruit is fine to buy for immediate use, but why it will not keep.

At the dinner table, try to describe what a carrot tastes like, or play a guessing game in which the person who is "it" describes an item and the others at the table guess what it is. The person who is "it" may also give hints to other family members—children usually begin with simple descriptions of color or shape, but soon concepts like acidity, where the item is naturally found, and who else eats this thing are added to the repertoire.

Observation and classification come naturally; the parent's goal is to look for those moments when children can be congratulated or joined in their observation and classification games.

No child is too young for classification games. "How is it," I say to my two-year-old, "that I know you're a person and not a cat?"

More often than not at this point she turns into a cat, but the fact that she knows cats crawl on all fours, say "meow," and lick up dinner tells me that she has many facts of cat behavior already in mind. If I, too, become a cat, I might even pass on a little information: "Meow, do you have anything to eat? I don't like lettuce, meow, do you have any fish? Have you seen our distant cousin the lion?"

Science in this sense gives you, the adult, a way to engage children in conversation and play.

Predicting, Estimating, and Measuring

You cannot directly teach someone how to predict, but there are two indirect ways to help develop this skill. The first is by focusing on natural opportunities: "How many stairs are there between the first and second floors?" "How many days before that rose will bloom?" "How long before the water turns to ice?"

You can also help develop this important skill by designing situations in which prediction is necessary. "Can you guess which of these two containers holds more water?" "Which holds more Ping-Pong balls?" "Which paper airplane will fall to the ground first—the one that's shot into the air or the one that's simply dropped?"

Virtually all of these estimating and predicting questions involve mathematics. This is no accident. Math is, in fact, a crucial language for scientists. Precise descriptions are best given in verifiable, replicable, mathematical terms.

Mathematics training in this country continues to focus on calculation. Addition, subtraction, multiplication, and division using cardinal numbers are taught in every school. Sadly lacking, however, is a concern with problem solving or relative value. It is important for children to be able to conjure up an image when someone says "3 grams" or "20 inches." In informal situations give as much attention as possible to mathematical descriptions; instead of telling Ben to put the rice into the pot, call it a cup and a half of rice, and help him notice that you are using twice as much water as rice when you cook it.

I recently sat in on a storytelling session that featured Leo Lionni's *Inch by Inch.* The book tells of an inchworm who escapes being eaten by a bird by offering to measure the length of the bird's song. The bird belts out his tune as the little worm crawls off "inch by inch." Most of the discussion, as one might expect, focused on how the worm had outwitted the bird. One child hesitantly raised his hand: "Couldn't you really measure a song by seeing how many inches you'd have to go before you couldn't hear the song anymore?" An adult who appreciated the scientific urge to measure and compare would take off from this remark by helping to organize a lively discussion or an on-site experiment.

Numerical relationships should be made meaningful to children. Baby sister Julie, for instance, can be described as weighing as much as two chickens. The distance before reaching a destination might well be described as four times the distance between our house and downtown. Concepts of "more" and "less" are also important to both children and scientists.

Manipulating Variables

Jean Piaget put it this way: "The goal of education is not to increase knowledge, but to create possibilities for a child to invent and discover." As children experiment, they come to see themselves as capable learners, interested in discovery and delighted with the power of their own minds.

Matthew had been playing with a balance, off and on, for about two weeks. His standard procedure was to put something on one side, it really didn't matter what, then grab another object and put it on the other. He would then announce which object was heavier, add something to the lighter side, and again announce which substances weighed more.

Recognizing the pattern, Matthew's babysitter put out two identical empty vials, which allowed the eight-year-old finally to have two things that obviously weighed the same. Matthew balanced the scale and the babysitter was pleased.

The next day Matthew arrived at the babysitter's very excited. Before going to bed an idea had come to him: the vials could be used to see which weighed more, salt or water. The babysitter watched as Matthew filled the two containers to the top. As he suspected, the salt weighed more. At this point the babysitter asked an honest question: "How much water would weigh the same as how much salt?" It took careful work to figure out the answer to that one, work that helped Matthew think of himself as a creative thinker, capable of answering even an adult's question. Experiments can also be used as a way to verify theories. When Nancy suggests that birds would prefer pizza to sunflower seeds, don't deny her assertions. Instead, help her design an experiment to test her hypothesis.

Naive Theorizing

David likes to figure out *why*. "I know why boys are stronger than girls," he begins.

"Why's that?" replies his encouraging but skeptical mom.

"It's like this," David goes on. "Boys have penises that stick out, while girls' penises are tucked in. That's why they don't have as much room for their strength."

Gulp quietly and think for a moment about David's explanation. The point of his statement, in adult terms, is that strength equals volume. And in many instances this is true—his father is bigger than his mother, and is also stronger. And David is quick to point out that he is stronger than his friend Katy, who is about the same size.

If you believe that it is important for your child to feel good about himself and his observational and analytic abilities, and yet you don't want him to think that you agree with a fairly outrageous explanation, what do you do at a moment like this?

Your first step should be to recognize your child's argument as a nice bit of thinking. The next step is to analyze carefully the content of the remark. In this instance, David is not trying to explain differences in male and female anatomy. (As a matter of fact, his explanation assumes that males and females have exactly the same equipment—just arranged differently.) The response of an adult, then, should address the issue of strength, although I would certainly make a mental note to discuss physiology sometime in the near future.

Your purpose then is to get the child to think a little harder. To indicate that you think his ideas are worthy of attention, begin with some examples that fit into the proposed theory: "That's why Grandpa is stronger then Grandma; I never thought of it that way." And then, as if you were thinking aloud, you might consider an example that doesn't fit as neatly, perhaps the case of two children who are about the same size, but here the girl is recognizably stronger.

David thinks for a moment. At this point he can either abandon or modify his theory, or he can fudge his data. Good scientists, and children who are learning to do science well, must accurately report what they find. Let them know that in a poem or story the author is free to manipulate characters as he or she sees fit, but scientists do not have this leeway.

So what do you do if David fudges his data and claims that the boy in question is stronger than the girl?

I recommend giving David another try. "What do you mean by stronger? Being able to ride a bike uphill? Being able to lift a heavy box more easily?" Or, depending on the way you are used to talking with one another, you might try a playful challenge: "You think Bob's stronger than Hillary—you've got to be kidding." What you are looking for is a way to allow your child to maintain his dignity and pleasing curiosity. An inquisitive, intelligent child who is not threatened will eventually worry counter-examples into place. In situations where the stakes are threateningly high, both grown-ups and children are more likely to distort information or alter their findings.

Integrating Inquiry

Inquiry skills needn't be taught as units. For children, this kind of breaking things into discreet parts is foreign and unsettling. Mia's father provides a perfect example of holistic science as he helps his daughter make scones. Together they measure out the ingredients and mix the batter. As they pat the dough into a circle, cut it into wedges, and brush the tops with beaten egg whites, Mia asks, "What's the egg for?"

"Good question," says her Dad. "Let's see what happens when we don't brush some with the egg."

The experiment is a grand success. They brush and they bake and they eat their results with gusto. The list of scientific skills learned from this pleasant and enjoyable encounter is substantial, although at no point did Mia's dad think of them as "objectives."

Mia's experience was in many ways more like that of a working scientist than the schemes passed off as the "scientific method" in schools. The basic question with which the child began was of the "what would happen if . . ." variety. She had no hypothesis and was not able to predict the outcome, although the next time she bakes, she may feel encouraged to brush the bread dough with egg and accurately predict what will happen under these circumstances.

Although most science educators would focus on the child's cognitive achievements, I am also pleased with the unintentional messages passed on through such an experience. Mia learned that science and experimentation are fun, and she is looking forward to her next scientific encounter. She is also proud that her father took her question about not brushing the scones seriously—seriously enough to design an experiment around it. Finally, the child learned that there can be an aesthetic and sensual dimension to science, and that knowledge does not detract from, in fact it enhances, an otherwise pleasant experience.

The scones event should make clear the potential of science as a family activity. Although a five-year-old could be challenged by the experimental nature of the project, a younger sibling could easily concentrate on the patting and pulling and tasting of dough. Furthermore, a nine- or ten-year-old in the family might have been able to predict the result of brushing the scones with various other solutions—"Would adding salt to the egg have made a difference?"—and to generalize and explain the information gathered. Making scones is possible in schools, but being able to repeat an experiment on several occasions and to appreciate it on a variety of levels is a pleasure most frequently reserved for home.

TEN DO'S AND DON'TS

1. *Build on children's questions.* Questions invite, in John Dewey's terms, *continuity* and *interaction.* That is, they connect with other interesting ideas and resources, and they encourage active experimentation. Look for ways to help children follow through on their own queries.

2. *Don't mistake reinforcement for real learning.* Just because you can ask Emily, "Do you know what this is?" and she correctly answers, "An orange grove" doesn't mean that she has learned anything. Open-ended questions such as "Which other fruits are in the same family as the orange? What do they have in common?" are generally more thought-provoking.

3. *Respect logic.* In science it is not enough to be right. One also has to be able to analyze and describe events and occurrences. Logic is the tool we have for judging the acceptability of any theory, procedure, or answer. So the next time you tell Abby that she can't have candy because it will rot her teeth, and she answers that she will go to the sink immediately after eating this Chompo bar, brush thoroughly, and use disclosing fluid to check for additional plaque, please admire her for her sound reasoning.

4. *Don't rescue a child from a difficult problem too early.* It's not easy to sit back and watch someone do something wrong. As I write this my small son is driving me crazy as he tries again and again to roll a piece of paper into the typewriter. First he had the paper in sideways, then he rolled it backwards. "I can't do it," he says, just as the paper magically slides into position around the roller. He types two characters, then, to my dismay, pulls the paper out of the machine to show me the fruits of his labor. I wince when I think of him trying to get the paper back in. But this time it goes better. He has not only learned how to get the paper into the typewriter, but also that he is capable of doing more than he thought possible. As the paper

goes in easily for character number three, he smiles over his shoulder . . . the world is his.

5. *Encourage the search for meaningful data.* Kristen wants to know more about the tentacles pea vines use to cling to a fence. She lists the factors she thinks might be important: time of day, the material being clung to (metal, wood, plastic), temperature and humidity. While in the garden she decides that the age and health of a plant might also affect "clingability." She purposefully manipulates conditions and keeps careful notes. Her thoughtful, direct approach is worth emulating.

6. *Don't underestimate the importance of intuition.* As a society we tend to overvalue those things that we can teach and measure and to undervalue that which we don't understand. Science and technology, we are told (and so we tell others), is a systematic study with a carefully defined method, best learned through careful plodding. But often people who are talented in science know something—that a given experiment will work, that one approach is less complicated than another—before they can explain why. It is true that these intuitive ideas have to be explained or worked out using standard scientific procedures and methods of verification. But intuition is an important attribute, one worth cultivating, practicing, and praising.

7. *Support intellectual efforts.* You are saying something important to Rini when you change the dinner menu from a five-hour stew to scrambled eggs so that she can use the kitchen as a laboratory. Justin, a space science enthusiast, appreciates the magazine with the latest photographs of Saturn that you pick up for him on the way home from work.

8. *Model scientific interest.* If your children see you reading *SCIENCE 86,* hear you discussing the geothermal changes caused by the eruption of Mount St. Helens, or watch you tinkering with natural dyes, they learn that science is an ordinary, but interesting, part of everyday life.

9. *Stress the interdependence of systems.* Everything that happens in the natural world affects and is affected by other physical phenomena. Children need to understand that any action or activity has consequences. For instance, every time we turn on the faucet we use water that comes from a res-

ervoir or well that has its own relationship to rain and clouds and condensation. As this water gushes down the drain and appears to leave our lives, we should remind ourselves and our children that it really hasn't left at all.

10. *Foster socially responsible science.* Science and technology are often perceived as existing within a vacuum. People actively engaged in these fields tend to see their work as pure, and its application as in the hands of others. On the other hand, those not involved in science feel they are ill-equipped to judge the wisdom of science-related decisions. In fact, we must all take responsibility for that which happens on and to our planet. Even young children, for example, can discuss pollution as a part of ecology and see themselves as active participants in making this world the kind of place they would like it to be.

DEMONSTRATING AND DISCUSSING VALUES

Bess-Gene Holt in her book, *Science With Young Children* (National Association for the Education of Young Children, Washington, DC, 1977), flatly states that food should not be used for play. That's right. Ms. Holt, a thoughtful and knowledgeable educator, is appalled by the macaroni necklaces, the seed and bean collages, and play dough made at home from flour, water, salt, and oil. Children, she contends, should understand that they live in a world where thousands go hungry, and that food, the very stuff with which Americans play, is what others need to fill their empty stomachs.

Children entering a class where this philosophy is practiced would be startled into thinking long and hard about activities they had taken part in without reflection. Counter-arguments could be made to justify "food play"; for example, more natural resources are used in the manufacture of plastic chips than would be wasted using plain dried beans and seeds.

Whichever position you support, the point is that science doesn't tell people what to do, but it does provide information that is useful in deciding how to live. Children learn this in two ways: first, by listening to adults discuss reasons behind their actions, and second, by noticing that people do things differently, and then asking why.

Every day, in a hundred small ways, you evidence your concern (or lack thereof) for the planet. If you tell your children why you recycle paper, why you carpool, why you buy energy-saving appliances, why you contribute to science-related organizations, you help them understand that there is a relationship between human knowledge and human values. Discussions about values also suggest to children that they, and people like them, can make a difference in what happens to the earth.

Science can also be important in a more personal sense. Tell children why you purchase or avoid foods containing nitrates, why vaccinations are important, why Grandma watches her sugar intake. Egocentric little beings like nothing better than understanding what makes them, and the people they love, function better.

Finally, it is never too soon to help children understand that science cannot provide solutions to all problems. At best, scientists can give descriptions and statistical probabilities, but science will never absolutely answer questions like "When does life begin?" or "Is it safe to live near a chemical plant?" To generate discussions that make this point, choose examples children are likely to understand. Science can tell them that this is a very rare wildflower, but they must decide finally whether or not to pick it.

Teaching Scientific Responsibility: Two Concrete Proposals

1. As soon as children are old enough to understand basic science policy decisions, they should be encouraged to write regularly to government officials, heads of companies, lobbyists, and citizen action groups, explaining their support, or lack thereof, for particular programs or policies. This is the world the children will inherit; they should tell powerful people, in no uncertain terms, about their hopes and fears for the planet.

2. Children can make financial contributions to organizations that repre-

sent their interests and concerns. Why not give your children a sum of money to donate to organizations of their choice? Dues from club memberships or monies earned through fund-raisers can also be earmarked for donations. As children examine literature from, for example, the Planetary Society, an organization dedicated to the peaceful exploration of outer space, or from the Sierra Club, a group active in the ecology movement, they learn about the technological and scientific ideas discussed as well as the importance of supporting the causes they believe in.

4 Answering Questions

If four-year-old Liza asks what the letters E-X-I-T spell, her father is pleased to answer "exit" and explain what exit means. If, on the other hand, she asks "What makes really hot metal look white?" her dad gives an embarrassed shrug of the shoulders, makes up an answer, or diverts attention with a comment like "You have to be older to understand that" or "What does that white color remind you of?" In effect, this parent teaches his child to stop asking questions about science.

Most of us are not scientists, but even those who are, are usually subject specialists. Although my husband can provide a cogent explanation of why hot metal appears white, he has little knowledge about what makes a cardinal's feathers red. Everyone who wants to help his or her child pursue scientific explanations must find ways to locate information. I emphasize information here for several reasons. Inventing a wondrous story about birds who learned to blend with the colors of autumn to avoid a wicked witch's spell is fun, but does not answer the question factually and is not science. For something to be "scientific," it must be based on verifiable facts and completely open to revision. Other people, working independently, must also be able to confirm the scientist's findings; that is, science must be publicly replicable.

There are three excellent sources for gathering scientific information: books, experts, and experimentation. As the child becomes more sophisticated, access and approach to these sources change.

BOOKS

To demonstrate the techniques used to find book information, I went to our *New Book of Knowledge* to answer the two questions posed earlier: "What makes a cardinal's feathers red?" and "Why does metal look white when it is hot?" I began by looking up "metals." Although there was some interesting discussion of melting ores and smelting metals, there was no reference to the color change I was interested in. From there, I reached for the "F" volume for "feather," and was told, through the index, to try "B," which I correctly assumed was for "bird." Here, too, I met with no success, but in reading over the entry noted some discussion of the function of feathers—some were lighter, some heavier—and there were allusions to protective coloration and the importance of color in certain forms of mating behavior. My inclination was simply to hypothesize about why the cardinal is red, but I recognized the importance of checking further.

While searching through the "bird" section an interesting thing happened: the children and I noticed several other fascinating topics, including a lively and totally unexpected section on prehistoric birds and clear directions on how to build a variety of birdhouses. After fifteen minutes of encyclopedia searching, I was not an inch closer to answering my two original questions.

This process could certainly be described as tedious and frustrating, especially comparing it to the easy satisfaction of identifying the word "exit" and giving its meaning. Yet in many ways, the knowledge gained from a search of this sort is commensurate with the effort exerted. The children learned about indexing; we formed an hypothesis which we discussed and could later confirm or deny; we realized some of the pleasures of informational browsing; and we discovered a project—building a birdhouse—that we were all anxious to try.

Our next stop was the library. I recommend that any family working together to develop their skills as scientists keep a list of questions and take it with them on a weekly trip to the library. At the library there are several sources worth checking as one pursues general science questions: science encyclopedias, books available on the shelves, books available through interlibrary loan, and periodicals or magazines. A game can be made out of this kind of search—one party can head for the card catalog while the other races to the reference section.

Asking the Right Questions

Part of what makes science searches difficult is that one needs to know enough to ask the right questions. The question about the cardinal, for instance, is not well defined. Do we want to know the chemistry of what makes something red on birds or are we interested in the cardinal's redness as a evolutionary adaptation? These topics would be housed in different sections of the library. Similarly, I might find a more helpful discussion on what happens when metal is heated in a book on molecular activity than in a description of the metal industry. Redefining and clarifying questions is, finally, a matter of practice.

The Search Continues

I began with the hot metal questions by looking in the *Raintree Illustrated Science Encyclopedia,* a book the librarian said was clearly written and often helpful. Nothing in the section called "metals" was on target, but I was struck by a mention that molecules were excited by heat. Could this account for the change in color? Next, I looked up "heat"; again there was nothing specific, but the list of related topics referred me to "color." Again nothing, but "color" referred me to "light." At this point I was frustrated with the reference work I was using and moved on to *The New Encyclopedia of Science* (Raintree, 1982).

Again I began: "metal"—nothing; "heat"—nothing specific, but there was a comment about stars. "Is the reason that metal glows white the same as the reason that stars are different colors?" I asked myself. This took me to a section on the absorption of light, where I was momentarily waylaid by fluorescence. Finally, in the section marked "light" I found it. Ah ha! Ah ha, ah ha! I said to myself proudly. And just to show you that I'm not faking:

Reference Books

Ideally, families should have on hand a good dictionary, an almanac, an atlas, and an encyclopedia. Reference sources make at-home questions easy, and give children some idea of where to begin with more complicated queries.

Dictionaries

Dictionaries help users define terms, and give brief, credible descriptions of words not understood. If, for instance, you've heard the term "trajectory," but aren't sure if it means any path taken by a moving object or that arclike movement taken by rockets shot into the sky, look it up. If, on the other hand, you want to know what breast cancer really is, a standard dictionary tells you nothing but the obvious.

An unabridged dictionary such as *Webster's Third International* plus supplement contains more than 450,000 definitions. By comparison, a collegiate or desk dictionary includes approximately 150,000 entries. Dictionaries designed especially for young people define still fewer terms, contain more illustrations, and generally use a sentence to demonstrate how words are used in context. Macmillan and Scott, Foresman produce children's and middle-school dictionaries that win high praise.

Almanacs

Almanacs, published annually, are designed to give current data and statistics about topics of general interest—population trends, weather, agricultural production, natural resources, and the like. Use an almanac to answer questions in which timeliness or a comparative analysis is important. For instance, an almanac is the perfect book to settle an argument about which spot is warmer: Tampa, Flagstaff, or Tel Aviv. Since all of the several almanacs currently available provide such data, opt for the one that is most graphically appealing and readily available.

Atlases

An atlas is a collection of maps; these may include maps of landforms, political maps, road maps, and thematic maps. Learning the geography of our planet is important for all educated citizens, and yet map study is one of these skills often relegated to a single social studies lesson. In the past few years several new children's atlases, both paper- and hardback editions, have appeared on the market. The well-respected *National Geographic Picture Atlas of Our World,* edited by Ross S. Bennett (National Geographic Society, 1979), is a personal favorite.

Encyclopedias

Encyclopedias are the most expensive and least understood family reference tool. They provide more detailed information than dictionaries and are also useful for people who like informational browsing. Electronics projects, woodworking ideas, and general science experiments are all in evidence. Encyclopedias should not, however, be used as a single authoritative work, nor as the major (or sole) source for a research paper.

A good new encyclopedia may cost $500 or more. [Each year the American Library Association reviews reference works and publishes their findings in a bulletin made available through public libraries. For children in grades 6 and below the *New Book of Knowledge* (Grolier), *The World Book* (World Book, Inc.), and the less comprehensive and less expensive *Childcraft How and Why Library* are highly regarded.] Used encyclopedias, however, have two major advantages: they can be purchased inexpensively and they teach children that the encyclopedia is not the place to be looking for current information.

To prove this point to children, several strategies work well. First, have them look up a topic such as space travel in a recently published encyclopedia, one published ten or fifteen years ago, and one from the turn of the century. As a child, I spent hours in similar searches, trying to make myself believe what was apparently true—that not everything written in a book is to be trusted.

It is also worthwhile to have children compare the depth of information offered in an encyclopedic account and that presented in a book. *Volcanoes in Our Solar System* by G. Jeffrey Taylor (Dodd Mead, 1983) provides the kind of detail an encyclopedia, by its very nature, is not capable of giving. Finally, I have children look through a book such as *Feral* (Macmillan, 1983), Laurence Pringle's account of animals gone wild, and try to find any encyclopedia category that would give them a handle on the questions Pringle asks and answers. There is nothing of the sort.

An added note: Any child working on a school report should begin by listing his or her questions about a given subject, not by looking up the topic in an encyclopedia. In this way, the encyclopedia's usefulness will be determined by the child's research interest, rather than by the amount of possibly irrelevant information it provides.

> As a glowing body gets hotter, it sends out more light of all wave lengths. But an increasing part of it is short-wave light. At 3000 degrees C most of the light is infrared. At 4000 degrees C red is the main color. The light appears yellow-ish white in these cases. At 5000 degrees C, the sun's temperature, all colors are strongly represented. The light includes a large amount of ultra-violet. The mixture looks white to the human eye. (p. 1083)

All of a sudden I recognized part of my problem. The light on the metal wasn't just white; it had gone through a series of color changes as the molecules were excited. "And that's why molten metal sometimes appears red," I surmised.

Although this search took at least thirty minutes, the time had been well spent. I had learned several things that might come in handy at other moments. I had the satisfaction of finding the answer to a difficult problem. And I now knew that if I wanted to find out more about the particular set of color changes that first interested my daughter and me, I would look in the physics section of the library shelves. I also felt a change in my self-image: although science is not always easy, I was capable of learning what I wanted to know.

I confess that at several points I was ready to give up my search. I wasn't sure I'd ever find an answer, and the problem seemed too difficult. Now this statement comes from a fairly disciplined adult. Think how hard independent research must be for a young person. This is just to say that if you spend some time helping your child with science searches, it may challenge you both.

My scientist husband was not as successful in finding the answer to the question about bird feathers. There were several random comments in science encyclopedias about how bright colors and fancy plumage attract mates to one another for breeding, but there was no systematic explanation. In other books about birds, there was one description after another about how feathers are used to fly, but virtually nothing on color. Which takes me to my next suggestion: If you try and can't find the answer to a question, ask an expert.

EXPERTS

One of the best and most satisfying ways to answer a question is to ask someone who knows or someone who has a good chance of being able to find out. The expert source you choose should depend on the nature of the question, the temperament of the child questioner, and the resources available in your community. For instance, a seven-year-old reluctant to converse with strangers may choose to write to an expert. Less shy children might simply telephone and politely ask if the expert could help with this particular query. Expert advice can be sought from the following people and places:

- Teachers, college professors, and industrial employees
- Editors of professional journals
- Museum personnel
- Science hobbyists affiliated with local organizations, such as nature clubs or computer groups
- Children's magazines with an "ask the expert" column

In asking for help from experts there are a few rules to keep in mind. First, make sure that the child's question can be discussed in a short conversation. To ask a subject specialist if he or she can list all the pretty birds in the world is not a good use of anyone's time. If, on the other hand, the child wishes to identify a rock and the field guides have proven confusing, ask for help. One of the best science projects I have seen included correspondence between a young man interested in the chemistry of kiwi fruit and a biochemist from New Zealand who had made kiwi the subject of his life's work.

Second, you should give an expert some sense of what you have done to pursue the subject: "I am very confused because the books I have read say that light is both waves and particles—do they mean that light is waves of particles or do they mean something different?" or "Why is it that a Boston fern, according to the directions in my houseplant manual, requires a lot of sunlight when all the members of the fern family with which I'm familiar prefer a heavily shaded natural setting?"

And finally, if your expert proves less than helpful, do not let this discourage you from seeking an informed opinion elsewhere. On occasion experts may have forgotten how to talk to those less knowledgeable than themselves; they may be unable to answer a seemingly simple question in terms an untrained person can understand (for example, in order to understand the answer to the question on light you really need some understanding of quantum mechanics); or they may be that rare breed who is just not interested in talking with young people. Children need to be reminded that there are as many different kinds of adults as there are kids.

Unable to find the answer to my child's question on the cardinal's feathers, I called Jed Burtt, an expert ornithologist who teaches at Ohio Wesleyan University. He told me, in effect, that he didn't really know why cardinal's feathers were red, and that he had spent much of his academic career trying to answer just that kind of question. His best guess was that the color of cardinals probably has to do with their mating behavior—the red color contrasts well with the cardinal's environs.

"Why would contrast be an asset? Doesn't the color of most animals serve as camouflage?" I pressed.

Dr. Burtt's answers were admittedly speculative. "Cardinals, unlike most birds, stay paired all year long, and their bright plumage may have to do with that fact."

Although I wasn't able to glean a definitive answer from my expert, I was delighted with the interchange. I learned some interesting things about cardinals. I learned that experts don't know everything and that my small daughter had asked a question that was worthy of serious research time. My child also learned that knowledge is a human construction, and that the reason most bird books didn't address the issue of color is that authors choose to tell readers what is known, rather than focus attention on yet-undiscovered phenomena.

EXPERIMENTATION

There is yet another suggested way of answering questions: finding out through experimentation. Our questions about metals and bird feathers don't lend themselves to experiments at home, but here's one that does. Janice asks if the tap water the family uses is hard or soft. Although she could answer this question through library research, or by calling an expert from the water commission, she can also test her water through a series of experiments—looking for residue in the tea kettle and comparing the sudsing ability of her tap water to that of "soft" rainwater.

Some experiments are obvious; if Jon asks if tadpoles really do turn into frogs, he can watch and monitor the growth of the amphibian without much trouble. Other questions require some knowledge of science. Margaret asks why salt keeps a boiling egg from cracking, which is slightly incorrect; the salt actually makes the egg coagulate quickly if it does crack. If a parent knew this fact, he/she could construct a simple series of experiments by adding more or less salt to the water and studying the cracks formed through a hand lens. If, however, one assumed that the salt somehow kept the egg whole, one would never know to look for hairline fractures.

The point here is that experimentation is most effective when combined with fact and theory. Informal science should not be allowed to develop or reinforce misconceptions. Without a basic understanding of the science involved in a given experiment, one may not have ideas for a next step or be able to interpret results.

Questions are the loom on which science is woven. Without questions, information, observations, predictions, and hypotheses become nothing more than a tangled ball of yarn. And yet there are few things more frustrating than having question after question go unanswered. Without strategies and resources like those described above, children finally turn away from science.

5 Community Resources

People learn about science and technology through home experiments, discussions, and reading. But much of today's science and technology cannot easily be replicated or understood in this way. Wherever you live, there are excursions you can take to help your children grasp the scale and the complexity of modern research and engineering. These opportunities should not be missed.

PLACES TO VISIT

Before deciding what places to visit, begin by simply listing the science spots in your area. Do this with your children; it's a good exercise in brainstorming and may give you an inkling of the sites they find most appealing. The categories indicated below should prove helpful. After you have your list assembled, organize it for your own purposes. Develop a coding system to mark places that vaguely interest you and places that you are anxious to visit. Put a check next to interesting short trips, and a star next to places that require an all-day commitment.

Institutions Dedicated to Science Education or Research

In this section list local science and technology museums, natural history museums, industry-sponsored museums or exhibits, zoos, planetaria, aquaria, children's museums with science exhibits, and nature centers. You should also check with local colleges and universities. They often have library exhibit space, or the halls of the science departments may be lined with items which interest youngsters.

Weather stations, model farms, paint testing sites, national laboratories, department of transportation projects, and a police crime lab or hospital lab may all be available for visits.

(Relatively) Undisturbed Natural Settings

If you live near a river, an unusual rock formation, a desert, a beach, a valley characterized by morning mist, a beaver dam, a cave, jot these down.

Managed Resources

List dairy, beef, hog, fish, poultry, tree, vegetable, and other kinds of farms. Also include research installations designed to reestablish a natural landform, (a prairie, for example) or to reintroduce an endangered form of wildlife.

Everyday People and Places

Using a book such as Vicki Cobb's *Secret Life of Hardware: A Science Experiment Book* (Lippincott, 1982), the most ordinary places become wild instruments on which to play strange music. Look carefully the next time you visit a service station, a car wash, a fix-it shop, or see telephone repair people at work. *Ten Minute Field Trips* by Helen Ross Russell (Ferguson, 1973) provides excellent suggestions for helping eight- to twelve-year-olds learn more about their community.

People—piano tuners, veterinarians, oceanographers, lab technicians, forest rangers, florists, and artisans of all types—can tell you much about science and technology if you simply ask questions as you watch them work.

Industrial Sites

Include all manufacturing plants that provide public tours. Particularly recommended are visits to places that produce products kids recognize: a bakery, a bottling plant, a bicycle manufacturer, a pencil factory, a newspaper production room, or a precision instrument company.

Waste Disposal and Treatment Sites

Water treatment plants, dumps, sewers, scrap metal or plastic companies, and recycling centers all make interesting trips. Warn children, however, about the dangers of industrial waste sites. Explain that even with highly specialized costumes and equipment, workers and nearby residents need to worry about chemical and biological contamination; radioactive waste provides its own complicated and unresolved problems.

Fuel Sources

Consider the various ways that people get energy for heat, cooking, electricity, and transportation in your area. Visits to a hydroelectric plant, the coal-fired smokestacks of a local university, or a company specializing in solar energy systems could be fascinating.

Communications Systems

Radio and television stations, monitoring stations for communications satellites, telegraph and telephone companies, are all possible visiting sites.

Transportation Systems

Cars, trains, planes, boats, subways, buses, trolleys, horse-drawn wagons, and the places that keep them in working condition are all worth an excursion.

Note: If you are on vacation, think about visiting science spots en route. Paul Hoffman's *American Museum Guides: Sciences* (Collier, 1983), a "critical handbook of the finest collections in the United States," and *Exploring Science: A Guide to Contemporary Science and Technology Museums* (available from the Association of Science-Technology Centers, 1016 16th St., N.W., Washington, DC 20036) are both helpful guides. Tourist information centers also respond to specific requests from bird watchers, rock hounds, or astronomy buffs.

GROUP VS. FAMILY VISITS

After assembling your list, give some thought to which places would best be visited by a group and which you could enjoy more as a family. For instance, many manufacturers conduct tours for schools or youth organizations, but not for individuals. Similarly, you might invite a speaker from the Department of Natural Resources to attend a formal meeting, though it would not be appropriate to ask him or her to present the same slide show to your family.

Several distinct advantages, however, characterize uninstitutionalized science. First, individuals have access to people and places groups may not. A child, for instance, can visit a pet shop regularly, become friends with the manager, and develop an insider's view of the problems involved in animal care.

Second, children usually talk more openly with family members about their perceptions and observations than they do in a group. Thus, you have a better chance of understanding the kinds of questions, frustrations, and interests your child entertains.

Third, informal visits are usually less hurried and less chaotic than group trips. An individual can spend hours watching workers construct a bridge, while organized groups are simply not ready for this kind of free-form observation.

And fourth, families are more willing to try the same spot again. Research indicates that a second or third sojourn may be extremely important, since children generally spend their first visit simply orienting themselves.

CHILD-SPONSORED OUTINGS

Children needn't be entirely dependent on their parents for science/technology-related outings. Share these ideas with your children, and see what adventures they can plan for themselves.

- One young girl, instead of selling lemonade, gave bug and bird tours of the two-block area surrounding her suburban neighborhood.
- Two other youngsters spent an entire summer marking the progress on an apartment under construction on a small model, adding to it each day as the construction workers added to their structure.
- A boy with artistic talents documented the developments in the vegeta-

ble garden on his sketch pad. Later he made a "flip book" which presented the flourishing of a zucchini in moving detail.

- Another child carefully marked a 2-foot-square segment of park. For a full year, she kept a diary describing the life forms that appeared there on Sundays.

GOING: PRACTICAL TIPS

One Saturday morning you decide to take the kids to the zoo. They whine about getting dressed and cry for more cartoons. Because you arrive late, the nearby parking lot is full. The baby's carriage won't open, your four-year-old spends the entire day chasing pigeons, the line for the toilets is long and smelly, and the only animal anyone remembers is the goat that ate the children's lunch. Yes, you are convinced that more science-related trips would be a good idea, but you need some inspiration and practical advice to get you going again.

- Children who do not often go places are much more anxious and resistant to excursions than regular visitors. Thus, if you find that within a four-week period you have not visited any of the places on your science list, try new strategies (see list below) to get yourself going.
- Plan several trips in advance. If you are the kind of person who does best with dates, times, and rainy-day alternatives, map out this kind of strategy. If, on the other hand, you are more spontaneous, keep ideas for trips simmering to celebrate specific events. For example, when Jessie finally wins a game of animal Yotta, take her to the zoo; be ready for Max's question about the rotation of the earth with a trip to the planetarium.
- Plan several short trips that you could take after school or in the early evening. Excursions need not be reserved for weekends and holidays.
- If you get impatient or bored being alone with your child, make arrange-

ments to take another adult along. Make it a very reliable adult, however; the last thing you want to do is cancel a trip because *your* friend can't come.

- If time is a problem, trade trips with friends—they take your child somewhere interesting one week, you take theirs the next. This is also a great way to share expertise.
- If you are reluctant to visit a spot because you know nothing about what goes on there, take a relevant book out of the library.
- Sign up for guided tours. This commitment should help get you to a spot at a particular time, and the tour will provide enough information so that, even if your child can't understand all that is being said, you can explain essential matters.
- Learn as much as you can about a place before you get there. If the food is notoriously bad, pack a lunch; if there are no changing facilities for the baby, bring along a rubberized folding mat so you can change him or her on the floor.
- Time your trips for less-crowded hours. If you anticipate long lines, bring along a book or drawing pad for the kids to use as they wait.
- See that your sightseers are dressed in comfortable clothes and shoes.
- Keep a knapsack or travel bag ready for excursions. Add items you want for your next trip—a flower identification book, disposable containers of juice—as you think of them.

HOW TO NOTICE MORE

Before starting a science trip, while visiting a spot, and after leaving the place, there are things you can do to help your child notice and remember more. Do not attempt to try each of these suggestions every time—excursions should never become labored. These suggestions are only a repertoire from which to pick and choose, note especially the ideas you and your children find most appealing and productive.

Before

To prepare for even a short trip, try the following. Locate available literature. Write for catalogs, brochures, and maps. Read books. Then plan your route—what are your children most interested in seeing? How are the exhibits arranged? When are particular exhibits likely to be too crowded?

With or without this literature, specific expectations and questions arise. Before a zoo trip, for instance, children may want to know what zoo animals eat, or how polar bears can exist in the summer heat. Rather than answering these queries with a simple "yes," "no," or "I don't know," try expanding on them. What *do* the various animals eat? Which ones are safe to feed? Is there a difference in the tooth structure of a carnivore and a herbivore? The polar bear question may lead to still other issues: What other animals are indigenous to cold climates? Do these animals adapt to our temperatures, or does the zoo maintain a climate in which they are comfortable?

Bookish children and their parents can radically heighten their level of awareness through reading. Whether you are looking carefully at desert flora, witnessing the hatching of a chick, visiting an air and space center, or studying a sundial, there are books designed to help you see more. For information on how to search out wonders like these, see chapter 7.

As an example of the way children's literature can inform a visit, I offer this abbreviated list of animal-related books.

For Young Children

Goor, Ron and Nancy, *All Kinds of Feet* (Crowell, 1984).
Photographs and text tell how feet work well for the animals who use them.

Kitchen, Bert, *Animal Alphabet* (Dial, 1984).
Exquisite, detailed drawings of animals are cleverly incorporated into the upper-case letters that begin the animals' names.

Roosevelt, Michelle Chopin, *Zoo Animals* (Random House, 1983).
Large pictures of toddlers' favorite animals with one or two simple sentences of text ensure the appeal of this board book.

Smith, Linell, *Who's Who in the Zoo* (Oak Tree, 1981).

The read-aloud set is enchanted by the rhymes, photographs, and cast of twenty animal characters described here.

Yabuchi, Masayuki, *Animals Sleeping* (Philomel, 1983).
Exquisite lifelike illustrations show the nocturnal habits of six animals.

For Grade-Schoolers

Cutchins, Judy, and Ginny Johnston, *Are Those Animals Real?: How Museums Prepare Wildlife Exhibits* (Morrow, 1984).
Clear text and excellent photos help answer questions about the techniques and materials used in preparing museum exhibits.

Emberly, Joan, *Joan Emberly's Collection of Amazing Animal Facts* (Delacorte, 1983).
Over 1500 uncommon facts about animals to fascinate and delight the curious.

Gustafson, Anita, *Some Feet Have Noses* (Lothrop, Lee & Shepard, 1983).
Know whose feet serve as spoons? This lively and useful work examines the types and functions of human, animal, and insect feet.

Hewett, Joan, *Watching Them Grow: Inside a Zoo Nursery* (Little, Brown, 1979).
Irresistible babies are cared for at the San Diego Zoo—this book tells you how.

For Middle-Schoolers

O'Connor, Karen, *Maybe You Belong in the Zoo* (Dodd, Mead, 1982).
Case histories of zoo and aquarium workers add interest to this thorough examination of zoo careers.

Shuttlesworth, Dorothy, *Zoos in the Making* (Dutton, 1977).
A lively discussion of solutions to typical zoo problems—animal care and propagation, exhibition design, and park arrangement.

Tongren, Sally, *What's for Lunch: Animal Feeding at the Zoo* (GMG, 1981).
Although this book focuses on the National Zoo in Washington, DC, the issue of health maintenance and the career oportunities described are relevant to all animal parks.

During Before entering a science spot, take a few minutes to study maps or brochures; decide what you want to see first, second, and what exhibits are less than appealing. If you wish to attend special presentations, get tickets early and schedule your visits around these events.

There are an increasing number of science and technology museums which invite visitor participation. In some places this is done through thought-provoking labels on displays; other institutions use technology—computer question-and-answer series or push-button controls. The dedicated hands-on museums insist on involvement: visitors literally sit on a centrifugal-force gadget, handle prisms, and manipulate communication devices. Places like these are living reminders that learning is an active, not a passive, process. If your child is working to understand what happens when a pendulum swings, don't denigrate his or her attempt by rushing on, claiming the subject is too difficult, or mocking these efforts.

Exhibits such as those described above are designed to provoke thoughtful questions and answers. If, after reading the documentation, you need help in understanding some aspect of a display, ask if someone is available to help answer questions. The Exploratorium in San Francisco, a model participatory museum, has "explainers" dressed in clearly identifiable outfits to help with such queries. Other institutions use docents (volunteer guides). Take advantage of the available resources.

- While your child explores and examines a science spot, encourage talking: "Does that creature remind you of any other beast?" "What would you like to know about that fish?" If you're playing with a magnetic device, think together about other applications of the principles demonstrated.
- Bend down and look at the world from a child's point of view. Discuss the differences in the way things look from your natural position and his or hers.
- If this is the second or third time that you have returned to a place, encourage your children to bring a friend. As they explain what they see to their guests, they, too, learn.

- Take books with you and compare artists' renditions or writers' descriptions to "the real thing."
- Try a thematic approach. Make this "ear" or "nose" day at the zoo, and notice the variations in form. When visiting the aquarium, look at the variety of ways underwater creatures move. Check out wheel design as you make your way through a transportation museum.
- Comparisons are also useful. Try to pair a trip to a large industrial bakery with a visit to a small mom-and-pop bake shop. What's different? What's the same? Take the same trip to the arboretum in summer, spring, winter, and fall.
- And last but not least, help your children to see science as a human investigation. Point out which dinosaur bones were found, and which had to be cast in plaster. Ask if they find the write-ups next to the exhibits helpful. Note who contributed to what research effort, and when. Have youngsters take notes (mental notes can be written up later) for curators of museums, rangers who care for nature preserves, and highway departments who design the roadside stops you use. Ask questions and tell officials what aspects of their designs work and which do not.

After Discuss and keep track of what children have seen, what they liked, what they'd like to see next time, and what simply didn't make sense.

Encourage young people to draw what they see or remember. Even the most unartistic viewers find themselves noticing detail and proportion in a new way. Diaramas and clay replicas also work well.

Encourage children to experiment with a variety of materials. Are the stones used to build this bridge stronger than those found near the house? What members of a bridge are most important in making it stable?

If your child becomes interested in agricultural variation, help set up experiments similar to those seen at the research station. Join your children at dawn to listen and see if the same birds that live in the wildlife sanctuary can be heard at home.

Use stories, poems, plays, dance, and music to help recapitulate a visit.

Your children may want to start their own junior museum or zoo. Help

Institutional Memberships

There are many reasons to become a museum, zoo, or aquarium member:

- You are more likely to visit frequently, and in doing so to explore the facility in greater depth. Go to the zoo one day just to see the lions, or to read a story about Curious George near the monkey cage.
- Members are always notified about educational programs the institution offers; member discounts are usually available.
- Other benefits may include a members' lounge, members' parking, special member events such as parties or tours, reciprocal agreements with other similar institutions, and discounts in the gift shop (in itself often worth the price of membership).
- In addition, as a museum member you are in a better position to lobby for the kind of institution you desire. For example, I would like to see a large institution give visitors lists of recommended tours: a "tired feet" tour, a tour for young children, a "what's unique in this institution" tour, and so on. I would also like to see museums develop relevant reading lists.
- It should be clear that memberships are a great bargain. But even if they weren't, I'd suggest joining institutions you see a valuable community resources. Museums, zoos, planetaria, and aquaria may not shut down without your support, but people like you provide necessary funds to develop new exhibits and try innovative programs. Your dollars, in effect, go directly to science education.

them define a space and develop descriptive and instructive information for their visitors.

Although recapitulation is important, the form of the rendering should definitely not be forced—there is nothing less poetic *and* less scientific than being told that you have to write a poem about hydrogen gas. Make this review natural and enjoyable. And if your children simply want to pass an experience by, or wait several months before discussing the event, so be it. You are there to make opportunities available, not to turn pleasant events into programmed learning.

SCIENCE GROUPS

A variety of science and technology clubs meet regularly in communities across the United States and Canada. Computer groups, amateur radio clubs, and astronomy organizations, for instance, are generally pleased to have interested young people join their ranks. Many of these enthusiasts began their hobbies as children and have a clear, empathetic knowledge of what kids can know and do.

Encourage your children to join one of these groups. A great deal of practical information gets passed around in meetings, information that is hard to learn from books. Here is a good place for children to get a sense of how much they know or don't know about a particular subject and to determine a direction for further investigations. In addition, I like what happens when children work and think with adults. In fact, the divisions in our society based on age or rank do not serve the scientific community well.

Other science groups are designed especially for young people. JETS (Junior Engineering Technical Society, 345 East 47th St., New York, NY 10017) sponsors engineering and science clubs for junior- and senior-high-school students. Their newsletter, "Jets Report," provides much useful information on science programs and contests, scholarship and work opportunities, and a variety of mind games and news stories. JETS also sponsors a National Engineering Aptitude Search Test and TEAMS, an academic competition. If young people do not have a sponsoring agency, or cannot generate sufficient interest to form a science club, they can become members-at-large and receive "Jets Report" nine times a year for $4.

Manufacturers of science and technology–related equipment have also become involved in organizing clubs which use their products. Apple Computers is sponsoring school Apple Clubs, and Estes, the model rocket company, helps young enthusiasts join together to develop their expertise. For further information on efforts of this sort, see the catalog section of this book under the subject headings.

Adult-run organizations such as the Astronomical League (P. O. Box

12821, Tucson, AZ 85732) or the Entomology Society (Box 4104, Hyattsville, MD 20781) also offer help in setting up organizations for children who are more comfortable working with their peers.

Many science and technology museums sponsor science clubs as well. From the literature they send, these programs generally look excellent.

All these science groups are basically designed for those with clearly defined scientific and technological interests. Other youth organizations, however, such as Scouts, 4-H, and church or synagogue groups, can (and often do) include science-related activities in their programs. If you feel that your son's or daughter's clubs need "beefing up" in this area, offer to arrange for speakers or field trips like those suggested in the section on "Places to Visit." Also, activity guides and materials useful to clubs are listed in the catalog section of this book under specific subject headings.

Two short reference works—Jean Kujoth's *The Boys' and Girls' Book of Clubs and Organizations* (Prentice-Hall, 1975) and Judith Erickson's *Directory of American Youth Organizations* (Father Flanagan's Boys' Home, Boys' Town, NE 68010, 1983)—are designed especially for young people in search of a group.

SCIENCE CAMPS AND SUMMER PROGRAMS

If your child has been stimulated by a school science program, has an ongoing interest in a particular area of science, or wants to develop new interests, a summer program may be just the thing.

Day-camp programs sponsored by community recreation departments often focus on nature study. Scout, Y.M.C.A., and private camps may include a science component. Furthermore, you can voice your concerns about informal science education to any camp director—ask specifically about projects, trips, and programs that might foster your child's interests.

There are also a number of private "sleep-away" camps that advertise themselves as science-oriented. The American Camping Association, an accrediting agency, annually publishes a *Parent's Guide to Accredited Camps* (Bradford Woods, Martinsville, IN 46151). In the back of this volume, camps list their specialties; "science" and "computers" are both used as reference categories. Science, in this case, means nature study, biology, chemistry, physics, rocketry, and so forth.

Other subject-specific camps can be found in periodicals designed for amateurs—astronomy camps, for instance, advertise in *Sky & Telescope.*

Both the National Audubon Society and the National Wildlife Federation have established ecology camps for young people and their parents. Audubon family camp experiences are developed around a theme: in 1984 the Maine site focused on ornithology, the Wisconsin center featured nature and the arts, and another Wisconsin ecology program involved camping, canoeing, photography, and nature study. An active preschool program is available for little ones. For further information write The Audubon Society, 950 Third Ave., New York, NY 10022.

The National Wildlife Foundations' camping program, the Ranger Rick Wildlife Camp, is located in North Carolina's Blue Ridge Mountains. Two-week sessions are designed to develop a love of nature and to encourage outdoor skills and abilities. For further information on the programs for nine-to-twelve-year-olds or the special teen session, write: Ranger Rick Wildlife Camp, National Wildlife Federation, 1412 Sixteenth St., N.W., Washington, DC 20036.

The American Association for the Advancement of Science Project on the Handicapped in Science has developed a state-by-state listing of "Out of School Oportunities for Youth": 500 local and state programs which are happy to accept, but are not specifically designed for, handicapped youngsters. This publication, together with a booklet entitled "Within Reach," is available for $3.50 from Project on the Handicapped in Science, AAAS, 1776 Massachusetts Ave., N.W., Washington, DC 20036. Because each state lists only a few programs, this brochure may be of more use to libraries or schools than to individuals.

Museums, zoos, nature centers, and other organizations committed to informal science education have recently established a number of well-conceived summer sessions. Check with institutions of this kind in your area for details as early as possible; these programs fill up quickly.

Children see summer as a time for fun and relaxation and relish the opportunity to vary the pace they've established during the school year. Summer is a great time for science.

6 School Science Fairs and Competitions

SCIENCE FAIRS

Twenty years ago, science fairs were presented as an optional activity for those deemed scientifically talented. With the new thrust toward universal scientific literacy, however, most schools require children to enter such a competition at least once, and encourage participation at other times. Science fair projects are generally planned and carried out at home; even if you have had no interest in science education, at this moment your child's involvement will be infectious.

An Overview

Science fair participants are expected to design and carry out their own research projects. Perhaps because young people have had so little experience in just this kind of activity, many feel nervous and threatened about taking on such a project. This situation is exacerbated by a concern that someone the students don't know will be judging the entry. In addition, the science fair is one of the rare times in the school curriculum when all work, not just the best, is publicly displayed. Then there are the numerous upsetting stories that circulate in every school—the judges are unfair and don't listen to what you have to say, Jimmy had to spend $75 on his project, Mary's parents built her setup, and so on.

But if the debits column on science fairs is long, the assets column is longer still. Science fair projects are just the kind of work students should be doing in science. If young people are distressed by the prospect of independent research, the problem is with the curriculum, not the science fair. The issue of "difficult" judges can fruitfully be seen as a lesson in using outside evaluators—sometimes they are excellent, sometimes they are unclear in explaining their perceptions. Students need to take these "disinterested" evaluations seriously, but they also need to develop the confidence to recognize when a "professional" makes a mistake. The best part about outside evaluators, however, is that they usually bring with them an expertise and a serious regard for science too often lacking in schools. These judges are also helpful in that they do not know who the straight-A kids are, and the distinctions they confer are often without that kind of bias.

Judges' views sometimes appear strange to participants. While entrants usually regard their projects as performances, staged for and dismantled after the big show, judges see the occasion differently. Most have volunteered because they wish to encourage an interest in and regard for science, and they see their role, in some sense, as educators. While students are anxious to hear their "grade," these experts are often more interested in suggesting further experiments, additional readings, or a modification in research design. I have often wanted to give a science fair follow-up prize for the entrant who has continued researching his/her project most intelligently after the fair ended.

The award system used in science fairs is also interesting. Generally judges award ratings rather than prizes; there is no limit to the number of projects designated "superior" or "excellent." And, science fairs are unique in that they are one of the few school occasions that celebrate an academic moment, rather than an academic career. Win or lose, it makes no difference whatsoever in the participant's standing the following year. Finally, I like science fairs because they are not completely teacher-directed. The science fair entrant can appropriately be likened to a student writing for publication: scholarship and presentation are what count.

Developing a Science Project

The most difficult and important part of developing a science fair project is finding an interesting and "do-able" idea. Although there are several excellent books to help in that regard, none adequately address that blank that potential entrants face as they try to take on independent research. The problem is in many ways analogous to "writer's block," and the techniques used to make people less afraid of developing their ideas on paper can aptly be appropriated by parents and science teachers.

- Try to figure out what the entrant is afraid of, because fear is what finally paralyzes those who see a science fair as too hard. Is it that the young person's expectations for him/herself are too high or needlessly elaborate? Is it that the child has never considered this kind of problem solving before and needs a model? Is it that last year's project didn't go well and the youngster is reluctant to try again?
- Let brainstorming be the first step in developing a project. Generally students can think of *some* ideas, but they have no way of sorting out which are more interesting and workable than others. Begin by listing subjects that the entrant sees as appealing: (a) pollution, (b) why do mangoes taste so good? (c) do hydroponics really work? (d) space travel, (e) something from my chemistry set.
- Help the child look at this list and expand it even further: What family resources are available? If Grandfather owns a fabric store, the child could investigate the science of textiles; if Mother works as a dietician, she might have access to useful information or untested ideas on nutrition. What hobbies does the child have? A kid who has soaked stamps for a stamp collection, for instance, asks why the paper on which stamps are printed doesn't disintegrate in water, but the envelope paper does. Look at everyday activities; on the way to school, what does he/she pass that seems somehow intriguing—a drug company, a dumping site, a bird sanctuary, an area where nothing seems to grow.
- Be explicit and concrete in explaining the guidelines used to evaluate science fair projects. Note, for example, that simply presenting an idea from a chemistry set is less than desirable. Trendy ideas—for instance,

pollution, solar energy, or computers—are generally more difficult to present in an interesting format than ideas judges encounter less frequently. Look at the judges' guidelines further on in this chapter, and try to translate these to the child in terms of the project he/she is considering. As in writing, an authentic question will lead to better results than some abstract, clichéd idea.

- Help the child make large ideas small, and put the small ideas within context. For instance, what special aspect of pollution catches his/her attention? Is it a concern with air, water, noise, chemical dumping, or garbage collection? Then push a little further. If the young person answers that it is water pollution, try to be even more specific: perhaps fishing in the West River has become terrible, or the algae on Lake Tendota made swimming there last summer disgusting. You may even find a real question embedded in this vague category "pollution"; for example, does thermal pollution seriously affect water life, or is this more media hype?

  Smaller topics can be expanded: What about mangoes tasting good? Is there something in the proportion of sweet and sour that might be worth investigating? Can this ratio be determined? Can we look for other foods that contain that same proportion to see if they are equally appealing?
- Assuage fears by inviting over entrants to last year's science fair. In looking for such a guest, try to find someone whose project was inexpensive, idiosyncratic, and fun. Virtually all the how-to science fair books cite projects completed by accomplished and dedicated high-school students who have won international competitions. I would prefer to talk with students who had a question they wanted to answer, and took the opportunity of the science fair to pursue this query.
- Beware of the equipment-intensive project. In this era when kids too often believe that money buys success, many middle-school and junior-high-school students in particular want to take the opportunity of the science fair to get their parents to buy them something outlandish. Be assured that the more of the project the student makes, the better it will be rated. If microscopes, computers, or precision measuring devices are

needed, borrow them from the school. Coming from a wealthy family should not be an advantage in a science fair.

- Help identify a project that uses the entrant's abilities to an advantage. For instance, if the young person is good at tinkering, note which ideas invite tinkering. If the student regularly goes off in the direction of theory, consider which projects benefit by this kind of theoretical speculation.
- *Most important,* help the entrant frame the chosen subject in terms of a question that can be answered through experimentation. Before beginning a project, a child should be able to complete this sentence: "I want to do ——— to find out ———. My hypothesis (guess) is that ——— will happen."

Experience suggests that teachers, librarians, science store managers, museum personnel, local scientists, engineers, and technicians can all be called upon once the student has a workable idea, but it is not appropriate to go to these people *for* an idea. Be aggressive. Write to Ph.D. scientists, M.D. researchers, or important politicians; many of them will write back. *Note:* keep all correspondence and include it in your science fair write-up. Show judges that you were seeking information, not ideas.

The methods and mechanics of undertaking a science fair project have been covered in many other books. To date, the two best are Joel Beller's *So You Want to Do a Science Project* (Arco, 1982) and *Nuts and Bolts: A Matter of Fact Guide to Science Fair Projects* by Barry A. VanDeman and Ed McDonald (Science Man Press, T.S.M. Marketing, Inc., Harwood Heights, IL, 1980). The Beller book is a detailed discussion of exactly what goes into a successful effort. Don't get this one for a poor reader; much of the time that should go into experimentation would be spent in reading the text. *Nuts and Bolts,* on the other hand, is a concise and remarkably sophisticated outline of information which is useful to anyone planning to enter a project. Both of these books are available through most of the science catalog stores and should be in museum shops, book stores, and libraries.

Thomas Moorman's *How to Make Your Science Project Scientific* (Atheneum, 1974) does an excellent job of describing the function of experimentation. Moorman also handles important issues such as the "deceptiveness of common-sense," evaluation, and proof.

For a professional/teachers' perspective on the workings and benefits of science fairs, nothing is better than *Science Fairs and Projects* (National Science Teachers Association, 1984). The articles vary in content—science fairs for young children, special fairs for the learning disabled, organizational suggestions, and the assessment of goals are all covered here.

Note: After the science fair is over, all those excellent projects should not be dismantled and forgotten. Some schools have arranged for a traveling exhibit through the town's elementary schools. In other communities projects have been displayed in enclosed shopping centers. A local science museum, children's museum, or nature center might also be interested in such a display.

JUDGES' GUIDELINES FOR SCIENCE FAIR EVALUATION

Science fair judges generally use guidelines like those below.* They are looking for:

√ **Knowledge Achieved** (Age and grade level are considered here.)

Does the entrant understand the scientific terms he/she uses?
Has the entrant learned principles or merely acquired a technique?
Has the entrant learned something in doing this project above and beyond that taught in class?

√ **Effective Use of Scientific Method**

Does the student have a clear purpose in undertaking this project?
Is the entrant aware of other theories or approaches relevant to his/her project?

* Adapted with permission from Judging Criteria used by the Ohio Academy of Science, 445 King Ave., Columbus, Ohio 43201.

Is there evidence of library or experimental research?
Has this effort been sustained and documented?
Has the entrant observed any basic phenomena?
Has he/she analyzed these observations in a logical manner, and drawn valid conclusions?
Does the research have a control experiment? Are the experiments a consistent set of studies?

√ **Clarity of Expression**

Can the entrant explain the project concisely and answer questions well? (Memorized speeches are not to be trusted!)
Is the written documentation presented well? Does it appear to be the student's own work, in his/her own words? (Be sure to keep a research notebook and bring it to the science fair.)
Is the display material neat and clear? Has the student checked the display for spelling and grammar errors?
When technical terms or equations are used, are they accurate? (Planetary orbits are not oval, but elliptical.)

√ **Originality and Creativity**

Is the approach to the research problem interesting or unique?
Has the student developed the approach on his/her own? (This one is very important! Beware of overly helpful teachers or relatives.)
Has she/he a unique or interesting presentation of materials?
Has the student adapted a kit to an original problem to economize on time and effort? (For instance, a project in which a student used a variety of materials in a pre-fab ant farm, instead of the sand the manufacturers sent, made it to a state science fair. Be warned—if judges see that an experiment is from a kit, it will be scored lower.)
Has the entrant made ingenious use of available materials and handmade elements? Can the entrant explain the construction and function of any devices used? (It is a good idea to set up a photographic record of a device being built and being used.)

Remember: A successful science fair project does not have to come to a successful conclusion. Judges look for process, rather than product. Experienced scientists know that experiments don't always work.

Science judges also insist that animals used in experiments must be properly cared for. These procedures are taken directly from the "Code of Practice on Animals in Schools" established by the National Science Teachers Association.*

Experimental Studies

1. In biological procedures involving living organisms, *species of plants, bacteria, fungi, protozoa, worms, snails, insects and other invertebrate animals should be used wherever possible.* Their wide variety, ready availability, and simplicity of maintenance and subsequent disposal make them especially suitable for student work.

2. *Some sample plant, protozoan, and/or invertebrate projects* include: field studies and natural history (life cycle, incidence in nature, social structure, etc.); germination; genetics; reproduction; effect of light, temperature, other environmental factors, and hormones on growth and development; feeding behavior; nutritional requirements; circulation of nutrients to tissues; metabolism; water balance; excretion; movement; activity cycles and biological clocks; responses to gravity and light; perception to touch, humidity and vibration; learning and maze running; habituation; communication; pheromones; observations of food chains and interdependence of one species on another.

3. *No experimental procedures shall be attempted on mammals, birds, reptiles, amphibians, or fish that cause the animal pain* or distinct discomfort or that interfere with its health. As a rule of thumb, a student shall only undertake those procedures on vertebrate animals that could be done on humans without pain or hazard to health.

4. Students shall *not peform surgery* on vertebrate animals.

5. *Examples of non-painful, non-hazardous projects on some vertebrate species* (including, in some instances, human beings) include some already mentioned under item (2) and also: group behavior; normal growth and development; properties of

* Reproduced with permission from *Science and Children,* September 1980. Copyright © 1980 by the National Science Teachers Association, 1742 Connecticut Avenue, N.W., Washington, DC 20009.

hair; pulse rate and blood pressure; various normal animal behaviors such as grooming, and wall-seeking; reaction to novelty or alarm; nervous reflexes and conditioned responses; special senses (touch, hearing, taste, smell, and proprioceptive responses); and respiration. None of these projects requires infliction of pain or interference with normal health.

6. *Experimental procedures shall not involve* use of microorganisms which can cause diseases in man or animals, ionizing radiation, cancer-producing agents, or administration of alcohol or other harmful drugs or chemicals known to produce toxic or painful reactions or capable of producing birth defects.

7. *Behavioral studies should use only reward* (such as providing food) and not punishment (such as electric shock) in training programs. Food, when used as reward, shall not be withdrawn for periods longer than 12 hours.

8. *Diets deficient in essential nutrients are prohibited.*

9. If *bird embryos* are subjected to invasive or potentially damaging experimental manipulations, the embryo must be destroyed humanely two days prior to hatching. If normal embryos are to be hatched, satisfactory humane provisions must be made for the care of the young birds.

10. On rare occasion it may be appropriate to *pith a live frog* for an educational demonstration. Correct procedure is rapid and virtually painless, and the animal should never recover consciousness. However, if done incorrectly, this procedure can cause pain. The technique should be learned initially using dead animals. Pithing live animals should only be undertaken by a person knowledgeable in the technique.

For conservation reasons, efforts should be made to protect depleted animal species such as *Rana pipiens.* Similar educational objectives can frequently be achieved using alternative species, or pursuing alternative methods of study.

11. *Protocols* of extracurricular projects involving animals *should be reviewed in advance* of the start of work by a qualified adult supervisor. Preferably, extracurricular projects should be conducted in a suitable area in the school.

12. High school students may wish to take assistant positions with professional scientists working in established, USDA-registered research institutions.*

SCIENCE COMPETITIONS

For students who prefer a more structured format and a lighter touch, there are other opportunities to demonstrate scientific reasoning, intuition, and knowledge. A number of science contests have been organized, for instance, which invite contestants to seek ingenious solutions to specific scientific and engineering problems. Materials required for these contests are inexpensive and available to all contestants. Participants sometimes work individually, and sometimes in four- or five-member teams. They use library resources, trial and error, systematic experimentation, intuition, or any combination of skills that enables them to develop a workable solution. The best-known of these programs are:

The Olympics of the Mind, a program for elementary- and middle-school students organized to foster critical thinking and creative problem solving across the curriculum (science and engineering included). For further information write: Olympics of the Mind Association, Inc., P.O. Box 27, Glassboro, NJ 08028.

The *Science Olympiad,* an interscholastic contest to increase interest in science and improve the quality of science education. Each year approximately twenty-five events, such as Science Bowl, Paper Airplane Race, Rocks to Riches, Metric Estimation, and an Egg Drop Contest are selected. Current divisions include one group for grades 6 through 9, and another for grades 9 through 12. An elementary division, for grades 4 through 6, is being considered. This program is less expensive and more science-oriented than the Olympics of the Mind. For further information write: National Science Olympiad Tournament, 5955 Little Pine La., Rochester, MI 48064.

Teams is a contest sponsored by JETS (Junior Engineerinig Technical Society) in which junior- and senior-high students compete in biology, chemistry, engineering graphics, English, mathematics, and physics. JETS also provides guidelines for an Engineering Design Contest, which can be used with any group of children age ten and above and can be adapted to home use. Write: JETS, 345 East 47th St., New York, NY 10017.

Those interested in a *National Bridge Building Contest* should write to: National Bridge Building Contest, Professional Engineers of Colorado, 2751 Alcott St., No. 167, Denver, CO 80211. The Illinois Institute of Technology, located in Chicago, and Brookhaven National Laboratory on Long Island also sponsor similar events.

Other contests of interest to school-age children are generally publicized in either *Science and Children* or *Science Teacher,* the two publications of the National Science Teachers Association. Manufacturers of photographic equipment and computer firms often sponsor competitions for users of their products, while organizations lobbying for a particular science policy have also been known to organize essay contests.

7 Science and Children's Literature

THE PLEASURES OF SCIENCE LITERATURE

Most people don't think of science books when they go to the library or bookstore looking for a "good read." They're wrong. To make this point, and to prove that reviewers do not all focus on the same qualities in a given work, I have asked a number of people involved with science books to share with you their standards for evaluation and some of their favorite titles.

Artful Science

For Pat Manning, a children's librarian at the Eastchester, NY, Library, the best science books provide emotional as well as intellectual satisfaction.

A lot of science books are what I call "useful books," the kind you need to do a school report or set up a science project. They do their job, which is to impart information. They've got indexes and glossaries, and in this sense they work well.

There aren't as many of those really neat books, the ones with the magic in them that make you run out into the backyard at night and stare up into infinity to see what's out there. Or to look at the ground beneath your feet and sense dimly the slow, inexorable surge of magma at unimaginable pressure. Or to see a piece of coal and realize in a quick burst that this black shiny rock was once stuff green and growing under a younger sun, and was chewed on by maybe dinosaurs . . . and now, burning, is somehow releasing the sunpower from those millions of years ago. . . . Even when your mother calls you to help with the dishes and your train of thought gets derailed, forever after there is something about coal that isn't just metric tons and bituminous and anthracite. . . .

There aren't many authors who are themselves caught up in the magic, and who want to and *can* share it with the young. When I find these more than useful books—sometimes not really *useful* at all—they do make an impression and I want everyone to buy one so that (a) kids will find out things instead of being told things and (b) those authors will not get discouraged and turn their hands to lesser tasks and (c) their publishers will not decide there's no profit in intellectual curiosity and opt for "useful books" instead.

I also wish that the truly awful books—the ones that read as though the author literally had one foot nailed to the floor and would not be set free until he had written a book, complete with glossary and index, on the Christmas Island Musk Shrew—would never be bought by anyone (even if it was the only book in the world on Christmas Island Musk Shrews, and Christmas Island Musk Shrews had suddenly become the focal point of the entire fourth-grade curriculum!). If that happened often enough, maybe those publishers would get the message that the reader's literary and intellectual sensibilities should be considered.

These books have been recommended by Pat Manning.

Inspired and Inspiring Science Books

Adkins, Jan, *Inside: Seeing Beneath the Surface* (Walker, 1975).

A unique, personal book (as are all Mr. Adkins's forays) using our mind's eye to look at the insides of things, to envision new perspectives: an apple pie, for instance, as seen by an engineer and a cook. Don't miss the author's caveats to would-be copycats!

Blassingame, Wyatt, *The Little Killers: Fleas, Lice, Mosquitoes* (Putnam, 1975).

A fascinating look at changes in history, large and small, caused by diseases transmitted by fleas, lice, and mosquitoes. Bubonic plague decimated the population of medieval Europe; typhus killed more of Napoleon's army on the retreat from Moscow than either the Russians or the bitter winter cold; yellow fever caused Thomas Jefferson, James Madison, and James Monroe to camp out in the public lobby of a tavern rather than face Philadelphia in the grip of the disease. Written by a man whose curiosity about the topic is evident.

Branley, Franklyn M, *Dinosaurs, Asteroids, and Superstars: Why the Dinosaurs Disappeared* (Crowell, 1982).

Sixty-five million years ago, dinosaurs (and a lot of other things) disappeared rather suddenly from the face of the earth. Mr. Branley speculates, explores several theories, and admits nobody knows why. An attractive book with a well-organized and engaging text. An index and book list for further reading are included, but don't let that prejudice you.

Lauber, Patricia, *Journey to the Planets* (Crown, 1982).

A beautiful book, albeit in black and white, showing the planets and moons of our solar system. Ms. Lauber manages to capture the awe and wonder of all the data our tiny probes have sent back to us across the millions of miles of space, and to show that in all we have found, only our own Earth, glowing blue and white and tan in the blackness of infinity, is capable of keeping us alive.

McNulty, Faith, *How to Dig a Hole to the Other Side of the World* (Harper & Row, 1979).

A super trip for the younger reader, or for anyone who has ever contemplated a patch of dirt, shovel in hand, with a speculative eye to digging all the way to China, complacently disregarding global geography and internal geology. Marc Simont's illustrations complement perfectly the informative, conversational text. No index.

Rice, Paul, and Peter Mayle, *As Dead as a Dodo* (Godine, 1981).

A superb example of beautiful bookmaking, carrying in its elegant beige pages serious reflection about the extinction of so many species. Luminous illustrations by Shawn Rice, accurate text, pleasing typeface, and an eye-catching cover. A picturely work of non-fiction to be shared. No index.

St. John, Glory, *How to Count Like a Martian* (Walck, 1975).

Beginning with a mysterious message from Mars (in beeps), this book provides an introduction to a variety of counting systems: Egyptian, Babylonian, Mayan, Greek, Chinese, and Hindu, with interesting stopovers

dealing with the abacus and the binary language of computers—then back to those Martians out there. Simple, beautifully laid out in blue and red text, informative and thought-provoking.

Sattler, Helen Roney, and illustrated by Anthony Rao, *Dinosaurs of North America* (Lothrop, Lee & Shepard, 1981).

A handsome compendium of our very own dinosaurs, from the tiniest coelurosaurs to the giant sauropods, whose fossilized footprints are larger than washtubs. These "safe" monsters (you will never find one lurking in your closet after dark) are depicted in all their awesome variety—spikes, crests, armor-plate, teeth, and claws. Aesthetically pleasing, accurate, and soul-satisfying to the dinosaur-phile. A booklist for further reading, a reference list by location of North American dinosaur discoveries (helpful for planning a special vacation), and an index, so you can look up your favorites at once.

Scott, Jack Denton, and photographs by Ozzie Sweet, *Discovering the Mysterious Egret* (Harcourt Brace Jovanovich, 1978).

A pleasure to mind and eye, as are most books by this dynamic duo, this work describes the life of a recent immigrant from Africa via South America. No one knows why the cattle egret has come to the New World, but a previously unoccupied ecological niche means they are here to stay. Nature abhors a vacuum. No index.

Ms. Manning's last recommendation is not least, nor newest. She writes, "Do try a book that has inspired more questions and more answer-seeking than any other I can remember:"

Burton, Virginia Lee (yes, she is the author of *Mike Mulligan and His Steam Shovel*) *Life Story: The Story of Life on Earth From Its Beginning up to Now* (Houghton Mifflin, 1962).

Life is presented here in five acts and many scenes, with wonderful things happening offstage in the wings. From the beginnings of our solar system in our galaxy, to what bulbs are up to under the dead leaves and February snows, in great looping spirals Ms. Burton gives us *everything*. The text is deceptively simple; the concepts are not. No index.

Ms. Manning's list is not complete. She has left out lots of our favorites and lots of her own. But every one of the books she recommends includes a perspective and a voice that says to children, "This book is being written by someone who finds satisfaction in knowing, and wants to share that informed pleasure with you." That is what distinguishes an excellent science book from even the best textbook or encyclopedia.

Relevant Knowledge

Bernice Richter, librarian at the world-famous Museum of Science and Industry in Chicago, and compiler (with Duane Wenzel) of *The Museum of Science and Industry Basic List of Children's Science Books 1973–1984,* does not focus as intently as Ms. Manning on the emotional and intellectual connections good science books provide. Rather, she discusses science books as instrumental in the creation of "scientific literacy."

> To be scientifically literate persons must be able to understand, use and interpret information and deal responsibly with science-related social issues. This requires a functional comprehension of science concepts that results from interactive experience. That's why my list of favorites contains so many experiment books that give children concrete experiences. Good science books also promote critical thinking and help develop an appreciation for the methods of science.

Developing Scientific Literacy

The following are recommended by Bernice Richter.

Abruscato, Joe, and Jack Hassard, *The Whole Cosmos Catalog of Science Activities* (Scott, Foresman, 1977).

A science activity guide in the tradition of *The Whole Earth Catalog*. In addition to science experiments, this book includes creative arts activities, puzzles and learning games, and science biographies. Appropriate (and fun) for classroom or home use.

Aliki, *Mummies Made in Egypt* (Crowell, 1979).

Describes the techniques and reasons for mummification in ancient Egypt.

Aliki's handsome watercolor drawings were adapted from paintings and sculptures found in tombs, and add much to the simple text.

Burns, Marilyn, *I Hate Mathematics Book!* (Little, Brown, 1975).
Events, gags, experiments, and activities to change you from a mathematical weakling to a heavyweight. Part of the fine Brown Paper School Series, which includes *Blood and Guts, The Night Sky Book, Reasons for the Seasons, The Book of Where, Math for Smarty Pants, Only Human,* and *Beastly Neighbors.* Although some volumes suffer from lack of an index, they are exciting books to dip into for imaginative activities, or to read from cover to cover.

Cobb, Vicki, *Science Experiments You Can Eat* (Lippincott, 1972).
Experiments with food chemistry demonstrate scientific principles and result in an edible product. The use of ordinary materials from your kitchen makes the examination of solutions, suspensions, colloids, and emulsions accessible to readers.

Flanagan, Geraldine Lux, and Sean Morris, *Window into a Nest* (Houghton Mifflin, 1975).
By building a nest box and affixing to it a concealed camera, the authors were able to observe the private world of a family of birds. Words and pictures unite to provide a magical record of their day-to-day activities. Books about nature and animal life are among the most popular science books for children; this is truly one of the most informative, moving, and thought-provoking of its type.

Macaulay, David, *Underground* (Houghton Mifflin, 1976).
Text and detailed illustrations describe the subways, sewers, building foundations, telephone and power systems, columns, cables, pipes, tunnels, and other underground elements found beneath a large, modern city street corner. Macaulay takes the reader on a dramatic and unique tour through this hidden subterranean world, and helps us appreciate the intricate network that allows our cities to function.

Macmillan, Bruce, *Apples: How They Grow* (Houghton Mifflin, 1979).
In this stunning black-and-white photo essay we see how apples grow from dormant bud to ripe fruit. Two parallel texts, one for young children and another for older readers, provide descriptive details.

National Geographic Society, Special Publications and School Services Division, *How Things Work.*
Uses familiar objects such as a bicycle and hot air balloon to introduce physics concepts. This title is representative of the fine publications produced for young people by the National Geographic Society. Combines excellent color photographs with a clear, informative text at an extremely modest price.

Pringle, Laurence, *Death Is Natural* (Four Winds, 1977).
A simple, moving text describes how death is a natural part of creature existence and makes life possible for succeeding generations. The almost lyrical quality of Pringle's prose and the accompanying black-and-white photographs help this book work on a personal level.

Stein, Sara, *The Science Book* (Workman, 1979).
A feast for young science experimenters and project buffs, this book suggests ways to use familiar surroundings—the backyard, home, pets, siblings, food, one's own body—to examine principles and phenomena of biology, physics, and chemistry. Explanatory sketches, photographs, and diagrams abound.

The Playful Challenge

Good science books can be exciting and challenging. To understand more about how books come to be lively, I asked Vicki Cobb, author of many highly recommended, lively science books—including *Fuzz Does It* (Lippincott, 1982), *Bet You Can't: Science Impossibilities to Fool You* (Lothrop, Lee & Shepard, 1980), and *Supersuits* (Lippincott, 1975)—to talk about how she engages her readers.

Forty years ago, when the first science books for children were written, the challenge was to present so-called adult scientific subject matter simply enough for a child to understand without sacrificing accuracy. At that time common wisdom said that science was beyond a child's comprehension and that it would be difficult to communicate basic concepts to children. Well, a host of writers including Julius Schwartz, Harry Milgrom, the Schneiders, Jeanne Bendick and others proved that one could write science simply and accurately for children. That, now, is a given.

Today's writer of non-fiction trade books for children faces another problem. We must write for a generation who expects, indeed demands, to be entertained. If readers are to acquire new information it must be presented palatably, requiring a minimum of effort on the reader's part for comprehension. As an author I must be constantly sensitive to my reader. To this end, I've developed a few do's and don'ts for myself.

I do try to grab attention with a bang at the start of every work. This means meeting children where they are, with something they already know about presented in an unusual way, preferably with an emotional hook. For example, here are some opening sentences from two of my books:

"Bet you can't read this book without trying at least one trick. They look so-o-o easy, you won't be able to resist. . . ." (From *Bet You Can't.*)

"Want to smell something rotten? Take a deep breath by a garbage can. If it's rotten your nose knows." (From *Lots of Rot.*)

Whenever possible I try to demonstrate a point with hands-on evidence. I encourage my reader to be skeptical and check things out personally. After all, science is a way of knowing by interacting with nature, as much as it is a body of knowledge collected incrementally by this way of knowing. Too many science books present science as if it were art history as opposed to teaching how to draw.

I try to constantly reconnect my reader with everyday life and personal experience. In *Science Experiments You Can Eat,* I explain the science behind the caveats every cook knows. (A tough pie crust comes from developing too much gluten, the protein in flour that you want to develop when you make bread.)

The most significant "don't" I have for myself is: Don't overkill with too much information! I try to keep the number of facts and terms down to the minimum needed to understand a concept. There is nothing like too much too soon to turn today's kids off science. It is my wish that my books be the first of many a child reads on a subject, not the last one he or she ever cracks.

Think carefully about what makes a science book appealing to your children. Then work with, rather than against, their expectations.

FINDING BOOKS

Using the Library

The Catalog

The first and most obvious place to look for books is the catalog of your local library. Library books are listed not only by title and author; a subject catalog should also be on hand. The subject entries are sometimes in the same spot as the author and title entries, and sometimes in a different place. Often children's titles are entered in a separate catalog in the Children's Room. If you can't find the children's catalog, ask the librarian for help.

Library users often make two mistakes in using the subject catalog. First, they assume that subject headings are as complete as author and title listings, and second, they spend considerable time copying down each number they find in the subject catalog. Unless you see a title that looks exactly right, don't copy a specific call number. In other words, use the subject catalog as a compass pointing to a general direction, rather than as a phone book where your party may or may not be found. For instance, in looking up a topic such as the moon, you find entries marked 523.3, 559.1, and 629.454; these refer to the moon in general, the geology of the moon, and explorations of the moon. On a scratch pad simply note the general sections, rather than specific titles, then move to these three sections to browse and choose the best books for your purpose. In general, the 500 section contains science books, and the 600s, technology-related materials.

Remember that the books you find on the shelves of your particular library do not represent all books on a given subject. Libraries have limited budgets; librarians study reviews, think of the community's needs, and make decisions. In addition, it may take your library a long while to process new books. Unfortunately, if you want to purchase a book for a gift, by the time it has appeared on the library shelf, it may already be out of print.

Book-Finding Tools

There are many reference tools housed in the public library to which patrons have access. Both the *Subject Guide to Books in Print* and the *Subject Guide to Children's Books in Print* are in no way selective, although they do give you a sense of what is currently on the market. With a broad topic like "birds," however, indexers are sure to miss many titles that you might wish to know about. If you are looking for books on a more specific topic, for instance "robins," these reference tools are more useful.

A selective list of children's book titles is provided by the authors of *Children's Catalog;* in general these works have received positive reviews in library journals. Note that this list is not complete, although it is often helpful.

Because *Children's Catalog* is considered standard fare in public libraries, the books listed in it can often be found on the shelves, procured from a nearby library, or borrowed through inter-library loan. These are also books that have been professionally reviewed; thus, you can check how professionals compared them with other books on the same subject. Many of the titles listed in the more complete *Books in Print,* for instance, are designed for supermarket checkout lines, and thus are not reviewed in library-oriented journals.

Bibliographies of Science Books

For a parent looking for books on a particular topic of scientific or technological interest, there are four sources I hope your library sees as indispensable.

- The American Association for the Advancement of Science produces a volume called *The Best Science Books for Children,* compiled and edited by Kathryn Wolff, Joellen M. Fritsche, Elina N. Gross, and Gary T. Todd. (Make sure that you are working with an edition published no earlier than 1983.) In its pages are listed over 1200 science and math books that the A.A.A.S., an organization concerned with scientific accuracy as well as format, sees as inviting reading. Subject specialists have reviewed each title listed here. The indexing, notations on reading level, and the handiness of the single volume make this work well worth its $16 price tag.
- A second excellent guide is *Best Books for Children: Pre-School–Middle Grades,* a sourcebook edited by John T. Gillespie and Christine B. Gil-

Science Book Evaluation Checklist

Phyllis Marcuccio, editor of the National Science Teachers Association journal, *Science and Children,* has specific questions she uses to evaluate science books for young people. Ask yourself these same questions as you attempt to judge the adequacy of a particular book.

√ Who is the *author?* How reliable? What background? What reputation? (Look at the book jacket for clues.)
√ What about the *content?* How important is the topic to children? How in-depth is the presentation? Is the sequence of ideas logical? How and why are various parts presented? Is the material *accurate?*
√ What about the *illustrations?* Are they accurate and do they match the text? When needed, is there a key to the diagrams? Are the illustrations clear and not overdesigned?
√ Is the *format* readable and pleasant?
√ Is the *vocabulary* broad and appropriate? There is no need to avoid "big" words in science, as long as they are explained and used in context. Is the *glossary* helpful?
√ Does the author use a "getting involved" approach—by encouraging observation and comparison, for instance?
√ Does the text encourage "how thinking" as well as "why thinking"?
√ Do the nature books offer "life cycles" and some natural history of the organism? Do they avoid (as they should) anthropomorphism, animism, personification, and teleology?
√ If the book deals with physical science or technology, does it contain the related fundamental scientific laws?
√ Are the suggested activities safe for children? Feasible? Are children able to perform the related tasks?
√ Are biases evident? I do not approve of books that are racist or sexist or books which celebrate violence. Is controversy handled fairly?

As examples of fine science writing for children, Ms. Marcuccio mentions several of her favorite authors, many of whom have produced several excellent books. Don't be discouraged if some of their works are currently out of print. Borrow the books you can find from your library, and wait hopefully for the day others will be reissued. Favorites that more than meet the criteria cited above are books by: *Glenn O. Blough* (along with his co-author on occasion, *Jeanne Bendick*), *Seymour Simon, Dorothy Shuttlesworth, Robert McClung* (a nature writer par excellence), *Millicent Selsam, Herman* and *Nina Schneider,* and *Roy Gallant.*

bert (Bowker, 1985). Experience suggests that the reviews here are accurate, the age recommendations are appropriate, and subject headings work well. This publication includes about 2000 science and math titles.

- A third, excellent source, *Science Books for Children: Selections from BOOKLIST, 1976–1983,* edited by Denise Murcko Wilms (American Library Association, 1985), is a handy compendium of annotated citations useful for parents and professionals alike. Titles are all recommended and cover books accessible to those in kindergarten through junior high. If bookstores don't stock this item, order a copy direct from: Publishing Services, American Library Association, 50 East Huron Street, Chicago, IL 60611.
- The Museum of Science and Industry in Chicago has a spectacular annual science book fair; in order to help visitors record what they see, Bernice Richter and Duane Wenzel have compiled an annotated bibliography of their exhibits called *The Museum of Science and Industry Basic List of Children's Science Books 1973–1984,* which includes most of the science titles currently on the market. The list can now be purchased for about $10 directly from the Museum of Science and Industry, or from the American Library Association. In this compendium each annotation is followed by a list of places where the book was reviewed, and a rating AA through D; thus, you can check on what others have said before purchasing a given title.

Librarians or informed bookshop owners can also be extremely helpful.

What's New: Professional Reviewing Sources

For the person looking to keep up with what is excellent in science books for children, there are several sources well worth your attention. Library journals, those publications designed to help librarians choose books for children, generally describe and evaluate the content of a work, and comment on the appeal of the book for the readers for whom it is intended. There is no reason why parents or teachers should not read library journals.

Your local library probably has at least one (and possibly all) of the following publications that conscientiously review children's books: *Booklist,*

Bulletin of the Center for Children's Books, Horn Book, Kirkus Reviews, and *School Library Journal.* Reviews usually appear within three months of a book's publication. Please note: reviewers for these journals do not always agree on their evaluation of a particular text. Also, all books are not reviewed, although about 90% of the titles recommended here are. My favorite reviewing sources usually include signed reviews (initials will follow the blurb); in time, you may find that you share the perspective of one reviewer or one journal.

All of these sources evaluate children's books, not just science books. Furthermore, most do not include subject headings, so to find science books in these publications, sit down with the journal and flip through until a certain title attracts you for one reason or another. For people particularly interested in a given topic, this is not the place to look for appropriate titles. If, on the other hand, you simply want to keep up with what's new in children's books, these works make good reading.

The most important sources for science enthusiasts to follow are two library journals dedicated entirely to science literature. Five times a year the American Association for the Advancement of Science publishes a journal, *Science Books and Films,* which reviews books, films and slides; there is a specific section dedicated to children's book reviews. A.A.A.S. is an organization of scientists, and their reviewers always comment on the accuracy and currency of the books' contents. Reviews are signed by field specialists, and age recommendations are included.

The other professional journal, called *Appraisal,* is specifically designed to evaluate science books for children. Published three times a year by the Children's Science Book Review Committee, a nonprofit organization sponsored by the Department of Science and Mathematics Education of Boston University School of Education and the New England Roundtable of Children's Librarians, this journal has an interesting format. Each book cited is reviewed by both a children's librarian and a science specialist; the evaluations become particularly interesting when these parties don't agree. I have talked to many librarians and teachers who did not know of the existence of *Appraisal* or

Science Books and Films. Either (or both) of these publications would make a lovely and useful gift from a friend of the library.

Science Award Books: Don't Miss These

Another source for excellent science reading is an award list presented jointly by the Children's Book Council and the National Science Teachers Association entitled "Outstanding Science Trade Books for Children." This list can be found annually in the March issue of *Science and Children,* or individuals who include a stamped self-addressed envelope can obtain one by writing to the Children's Book Council, 67 Irving Pl., New York, NY 10003. Although criteria for selection appear fairly standard—books should be accurate, readable, and exhibit a pleasing format—I find that this is the best single source for new books of general scientific interest. Reviews are signed by the distinguished committee members, and regular readers learn to seek out books recommended by their favorite reviewers.

Other science reading awards also give direction to non-fiction enthusiasts. Each year the New York Academy of Sciences (2 East 63rd St., New York, NY 10021) honors several books with their commendation. These winners are listed in the *New York Times,* among other places. I am impressed with this group's choices.

The Washington, D.C., based Children's Book Guild annually salutes an author of non-fiction. This award is given for a writer's collected works rather than a specific text, and recent winners Tana Hoban, Patricia Lauber, Millicent Selsam, and Isaac Asimov are all worthy of your attention.

The Reading Teacher, a journal published by the International Reading Association, publishes a list called "Children's Choices for 19——," which includes the category "Informational Books." Children's choices should be valued. For their most current list write: International Reading Association, 800 Barksdale Rd., Newark, DE 19711.

Science Book Reviews Specifically for Parents

Beverly Kobrin, an experienced teacher with a Ph.D. in Education, publishes a newsletter that provides topical reviews of children's non-fiction. Although *The Kobrin Letter* does not include as many reviews as, for example, *Ap-*

praisal, the topics—for example, the zoo, desert life, and "for experimenters and puzzlers"—are those a parent or teacher is ready to use.

When Dr. Kobrin was asked to specify the criteria she uses in judging science books, and then to list several titles that meet those standards, she replied:

> I look for books that in addition to meeting the essential non-fiction criteria (accuracy, currency, a clear distinction between theory and fact, for example), satisfy curiosity but not sate it; inspire as many questions as they answer; and create an excitement and irresistible enthusiasm for the subject.

Dr. Kobrin goes on, "The following books once opened, cannot be left unread, and once read, will not be read just once."

Bester, Roger, *Guess What?* (Crown, 1980).
Black-and-white photographs and a few words provide clues to the identity of seven familiar animals.

Diagram Group, *Comparisons* (St. Martin's, 1980).
Hundreds of detailed illustrations that show how people, plants, animals, and objects compare in measurable attributes; rivals the *Guinness Book of World Records* in popularity with children.

Fischer-Nagel, Heiderose and Andreas, *A Kitten Is Born* (Putnam, 1983).
Simple text and color photographs record the birth and rearing of Tabitha's litter of three kittens.

Grillone, Lisa, and Joseph Gennaro. *Small Worlds Close Up* (Crown, 1978).
Black-and-white micrographs of animals, vegetables, and minerals magnified from 100 to 16,000 times.

Miller, Ron, and William K. Hartman. *The Grand Tour* (Workman, 1981).
Space traveler's guide to our solar system; vivid text and colorful illustrations.

Potte, Tony, *Robotics* (E.D.C., 1984).
Just one of many titles in a series of publications with well-written text in-

tegrated into colorful, cartoon-style, accurately drawn, detailed illustrations children cannot resist.

Taylor, Paula, *Kids Whole Future Catalog* (Random House, 1982).
View of today's children's adult world as projected by scientists in photographs and illustrations; sources for free or inexpensive related materials provided for each of 100 topics.

Tison, Annette, and Talus Taylor, *Adventures of Three Colors* (Merrill, 1980).
Pictures in primary colors printed on clear plastic allow children to create new pictures in mixed or complementary colors by flipping one page over the other.

Wexo, John, *Zoobooks.*
One-a-month series on wildlife education, these twenty-page booklets are distinguished by superb color photographs and illustrations and lucid writing.

If the suggestions on Beverly Kobrin's list appeal to you, try a subscription to her newsletter, published ten times annually. To order, send $12 to *The Kobrin Letter,* 732 N. Greer Rd., Palo Alto, CA 94303; for $2 you can get a sample issue.

Two other parent-directed newsletters carry extensive lists of non-fiction along with other material. *Parents' Choice,* a quarterly review of children's books, movies, television shows, records, toys, computer software, and video cassettes, is a substantial and interesting publication. A subscription is $10; for further information write: Parents' Choice Foundation, Box 185, Waban, ME 02168.

Selection, a glossy quarterly review for parents ($12 annually), uses a thematic format to discuss children's books and activities. Several issues have focused on science and technology. For more information write: P.O. Box 5068, Stanford, CA 94305.

I've saved a real favorite for last. Every December Philip and Phylis Morrison write a Christmas survey of children's books on science and technology for *Scientific American.* The Morrisons' annotations are detailed and accurate, they cover a wide variety of sciences, and the more than twenty books they suggest can all be found (or ordered from) your local bookstore in time for the holidays. This is a list which should not be missed.

Fiction

In looking over lists of recommended titles, it is clear that science books have been identified with non-fiction. From conversations with children, however, it is also clear that there are many who take their scientific interests to and from the fiction they read. Science fiction is a genre with tremendous appeal to science fans, especially when the included science seems plausible. The works of Isaac Asimov, Arthur C. Clarke, Larry Niven, and Poul Anderson are especially popular with this crowd.

Science fantasies may appear problematic to information-oriented adults. With America's fear of science comes an unreasonable belief in its possibilities. This propensity is exacerbated by the wonders of trick photography—as cinema and television heroes defy the laws of nature, what is possible and impossible becomes harder to sort out. Furthermore, parents not schooled in science may have trouble figuring out what is plausible and what is not. Three books, Amit Goswami's *The Cosmic Dancers: Exploring the Physics of Science Fiction* (Harper & Row, 1983), Neil Barron's *The Anatomy of Wonder: An Historical Survey and Critical Guide to the Best of Science Fiction* (Bowker, 1981), and Peter Nicholls and David Langford's *The Science in Science Fiction* (Knopf, 1983) written for an adult audience but accessible to advanced child readers, will help determine what is scientifically probable, possible, and impossible.

Realistic fiction also holds great promise for a reader who wishes to see science as played out in a story. In a book such as *Julie of the Wolves,* Jean Craighead George, herself an author of many excellent non-fiction titles, uses impeccable care in describing the conditions that allow for her heroine's survival. After reading *Winnie the Pooh* (or even before), try standing on a bridge and playing "pooh sticks." Even in a novel such as Laura Ingalls Wilder's *Farmer Boy,* one finds science experiments of a sort. Is Alphonso correct when he claims that you can fill one glass with popcorn and another with milk, then one by one you can put every popped kernel into the liquid without it spilling over?

Here is children's librarian Rachel Alexander's abbreviated list of fiction titles that include a good deal of science.

Recommended Fiction with Science

Cleaver, Bill and Vera, *Hazel Rye* (Harper & Row, 1983).
Eleven-year-old Hazel's lack of appreciation for the land and growing things is transformed as she watches an impoverished neighbor bring an orange grove to life.

George, Jean Craighead, *Julie of the Wolves* (Harper & Row, 1972).
Julie, a young Eskimo girl, survives a trek across frozen Alaska only because she's befriended by, and learns to live with, a pack of wolves.

Holling, Holling C., *Paddle-to-the-Sea* (Houghton Mifflin, 1941).
From a melting snowbank in Ontario, north of the Great Lakes, a young boy launches his hand-carved Indian-in-a-canoe. Richly colored illustrations, map inserts, and detailed diagrams help readers follow its journey to the Atlantic.

Hurwitz, Johanna, *Much Ado about Aldo* (Morrow, 1978).
All goes well with the classroom science project until eight-year-old Aldo realizes that the crickets he's come to know will soon become prey for the newly arrived chameleons.

Kent, Jack, *The Caterpillar and the Polliwog* (Prentice-Hall, 1982).
Cartoon-like illustrations and humorous dialogue help little people focus on the changes a caterpillar and polliwog make. Characters and readers together are awed by the excitement of everyday nature.

Mazer, Harry, *The Island Keeper* (Dell, 1981).
Desperate to prove herself, fat, rich Cleo runs away to a deserted island. She survives the summer months, as planned, but a cruel twist of fate forces the teenager to endure the brutal Canadian winter alone.

Miller, Edna, *Mousekin's Close Call* (and other Mousekin books) (Prentice-Hall, 1980).

When attacked by a swamp sparrow bent on protecting her young, a weasel drops Mousekin and runs. Mousekin responds by playing dead in this adventure story which relays accurate scientific information about instinctive behavior.

O'Dell, Scott, *Island of the Blue Dolphins* (Houghton Mifflin, 1960).

An Indian girl, served well by observation and invention, survives eighteen years alone on an island off the coast of California in the 1800s.

Rockwood, Joyce, *To Spoil the Sun* (Holt, Rinehart & Winston, 1976).

Lessons from nature abound in this moving historical re-creation of Cherokee life and the arrival of the whites, as seen through the eyes of a 16th-century Indian girl.

Speare, Elizabeth George, *Sign of the Beaver* (Houghton Mifflin, 1983).

Left alone to guard the family's wilderness home in Maine, an 18th-century Anglo boy is hard-pressed to survive until Indians teach him the ways of plants and animals.

Taylor, Theodore, *The Cay* (Doubleday, 1969; Avon, 1977).

A young blind American boy and old West Indian man, shipwrecked on a Caribbean cay, overcome distrust, prejudice, and physical hardships in order to survive.

Another kind of science-related fiction is typified by the *Einstein Anderson* series, written by noted science author Seymour Simon. These extremely popular collections of short stories focus on a sixth-grade boy who uses his scientific methods and knowledge to solve a variety of short mysteries. This series is especially recommended for elementary-school children who are taken with "The Bloodhound Gang" on *3-2-1 Contact.*

Biographies

There are a number of children who enjoy biographies of scientists. The problem here is that most biographies for children harbor strange notions of both science and human achievement. Moreover, many are simply inaccurate. In biographies of Marie Curie, for instance, one finds illustrations of the great scientist receiving her first Nobel Prize (when in fact she was in France suffering from radium poisoning); she is pictured in a variety of lovely dresses (in reality she owned only two frocks); and more to the point, she is regularly portrayed as a dreamer, looking dramatically into the beyond. Curie was, if anything, a disciplined, unfrivolous positivist who would surely have been distressed by this notion of her personality and her work. On the other hand, science is selectively presented as a subject for the brilliant plugger. Children are told that luck and inspiration rarely figure in the formula for success.

There are, however, several excellent biographies that might be read as models, and which can provide the basis for an interesting discussion about what a science biography should and should not do. Especially recommended are Vivian Grey's *The Chemist Who Lost His Head: The Story of Antoine Lavoisier* (Coward-McCann, 1982), and Nancy Veglahn's *Dance of the Planets: The Universe of Nicolaus Copernicus* (Coward, McCann & Geoghegan, 1979). In *Test-Tube Mysteries* by Gail Kay Haines (Dodd, Mead, 1982), definitely a good read, science is fittingly described as a kind of detective work. All of these books are intended for children in grade 6 or above. Valjean McLenighan's *Women and Science* (Raintree, 1980), appropriate for a slightly younger group, describes both the frustrations and the variety of satisfactions Florence Sabin, Chien Shlung Wu, Margaret Mead, Alice Hamilton, and others have encountered in their work.

Biographies are also a superb place to look for experiments that accord a child the dignity of a professional adult. Most early scientists worked with items less sophisticated than those available to children today. Particularly noteworthy are Galileo's exploits, the microscopy of Anton van Leeuwenhoek, and the nature studies of Cordelia Stanwood (see, for instance, Ada Graham's *Six Little Chickadees: A Scientist and Her Work with Birds* [Four Winds, 1982]). Biographies can also give readers some sense of the patience, persistence, and self-confidence (as opposed to expensive gadgetry) these scientists used to undertake their work.

Another form of biography which has received considerable play in the past few years focuses on the lives of dedicated men and women who have *not* become famous. Barbara Land's fascinating description of scientists in Antartica, *The New Explorers* (Dodd, Mead, 1981), William Jasperson's *A Day in the Life of a Marine Biologist* (Little, Brown, 1982), and Melvin Berger's *Disease Detectives* (Crowell, 1978), which describes the activities and history of the Center for Disease Control in Atlanta, Georgia, are all excellent examples.

Unfortunately, there do not appear to be many good biographies for young children. Alice and Martin Provensen's *The Glorious Flight: Across the Channel with Louis Bleriot* (Viking, 1983) is gorgeous, accurate, and accessible to younger children with an adult's help. *Veterinarian, Doctor for Your Pet* by Arline Strong (Atheneum, 1977) may work well for young ones already familiar with the profession. The now classic D'Aulaire biographies, especially the ones of Benjamin Franklin and Christopher Columbus, at least acquaint little people with the names and major achievements of these men, and introduce children to biography as a literary form.

WHERE CHILDREN AND BOOKS MEET

Most of us, readers and reviewers alike, look at a science book and ask, "Is it accurate?", "Is it readable?", "Are the proposed activities safe and enjoyable?" However, few people in our culture seem to ask if what a given book teaches is worth knowing, or what the logical consequences are of supporting a given scientific position. These kinds of questions always jolt me into rethinking my values and priorities. There is one reviewer, Lazar Goldberg, a professor of science education of Hofstra University, whose comments on a book invariably help me recall that larger picture. I conclude this chapter with Dr. Goldberg's statement on what he looks for in a science book. It is a perspective that gives meaning to all that you do with children and science.

> We educate children who will live most of their lives in a time we cannot describe with any degree of accuracy. What will continue to be useful knowledge irrespective

of the changes that are bound to take place? At the same time that we provide education of lasting value, we must provide experiences that are appropriate for children as we find them. Parents and teachers know that admonitions such as "When you are older you will be glad that you have learned . . ." are largely meaningless to children. Interesting and significant experiences that are right for children now are precisely the ones that will be most useful later. There are three principal categories of science books for children that fulfill these criteria.

The first of these categories includes books that stimulate children to look for the puzzling, the mysterious, the remarkable in commonplace experience. Such books foster questions and the isolation of problems that invite pursuit. How is it that rubber balls, glass marbles, plastic billiard balls, all bounce, but wet clay does not? How do we manage to blow cool air to relieve a burn and warm air to relieve chilled hands? Why do puddles freeze from the top down and not from the bottom up? What makes soap bubbles take a spherical shape? What pushes the water up to the faucet? How does the skin "know" to stop growing over a cut? Why is it harder to keep your balance on a bicycle when you ride slowly than when you ride rapidly? How does a fly manage to stand on the ceiling? What enables us to hear around a corner, but not see around a corner? When the lights are turned off, why is it dark? Where has the light gone? Why are there two holes in an electric socket? Why is a vacuum cleaner so noisy, and what makes it suck up dirt? And why is there still dust in the house after it has been cleaned thoroughly? Good science books invite children to observe carefully, to reflect on their observations, to ask questions and search for solutions any way they can.

Another category of science books helps children find meaning in the seemingly random, chaotic, disparate events in the sky, on the earth, in matter and energy, and in living systems. The sky appears to consist of a jumble of countless stars. The surface of the earth and whatever lies below is no less baffling. What sense is one to make of light, electricity, automobile engines, the myriad substances with which we are surrounded, and the vast scale from the infinitesimally small to the immeasurably large? How is one to comprehend the number and diversity of living things? It is uncomfortable to live in an apparently disorderly world. Good science books help children find intellectual and esthetic satisfaction in the search for and discovery of order.

A third category of science books helps children make connections, to relate their various experiences so that what they learn gives them ever increasing power to learn independently. They learn when it is appropriate, to count and measure, but that measurableness is not next to godliness. They learn that stars have histories and that history is essential to an understanding of living things. They learn that to do science one must communicate one's findings. They learn that there is a scientific component to the most pressing problems of the day: nuclear energy and the danger of nuclear war; environmental decay; hunger and poverty; racism and other forms of bigotry; population growth; the extinction of many species of plants and animals. They learn that knowing and feeling do not dwell in separate abodes, that scientists feel and that artists comprehend.

What follows is a small selection of the many books Dr. Goldberg sees as contributing to children's understanding of science.

Pringle, Laurence, *The Hidden World: Life Under a Rock* (Macmillan, 1977).

Rocks are found in so many places, both on land and in water. Under many of these rocks little worlds of plants and animals can be found. To enter these worlds one need only lift a rock.

Nestor, William P., *Into Winter* (Houghton Mifflin, 1982).

There are many things to investigate in winter. This book describes not only what the reader can investigate but how.

Gardner, Robert, and David Webster. *Shadow Science* (Doubleday, 1976).

Everyone can make shadows, but few realize how much you can find out by exploring them. You can tell time, find direction and the season, understand night and day, and more with the help of this book.

Faraday, Michael, *The Chemical History of a Candle* (Viking, 1960).

Although the great scientist presented these lectures to young people more than 120 years ago, the lesson about the many things we can learn from a careful examination of so common a thing as a candle is still most instructive.

Adler, Irving, *The Stars* (Crowell, rev. ed. 1980).

The author helps us find our way among the stars. What is more unusual in a book for young readers, he explains how astronomers know the things they know.

Leakey, Richard E., *Human Origins* (Lodestar, 1982).
The story of our history is beautifully told and illustrated. The author, an expert in this field, includes views that are different from his own.

Helfman, Elizabeth S., *Signs and Symbols of the Sun* (Houghton Mifflin, 1974).
People have been interested in the sun from many points of view. The sun has been prominent in religion and mythology, in science and art.

Weiss, Ann E., *The Nuclear Question* (Harcourt Brace Jovanovich, 1981).
Nuclear energy, like technology in general, is neither good nor bad. It is subject to social application. This principle is illustrated by examining the history, application, and problems of nuclear energy.

Maruki, Toshi, *Hiroshima No Pika* (Lothrop, Lee & Shepard, 1982).
A Japanese artist tells the story, largely through pictures, of what happened that fateful day, August 6, 1945.

CHILDREN'S BOOKS LISTED IN THIS CHAPTER

Adults looking for children's books generally use reading level as a guide, and in keeping with their expectations I have arranged the following titles accordingly. But because both people and books are hard to characterize neatly, there are many titles which could easily be put into two or more categories; these are noted with an extended-grade-level recommendation in parentheses. Also, remember that in informational books, interest is considerably more important than reading level.

For All Ages

As Dead As a Dodo, Paul Rice and Peter Mayle (Godine, 1981)
Comparisons, Diagram Group, (St. Martin's, 1980)
Dinosaurs of North America, Helen R. Sattler (Lothrop, Lee & Shepard, 1981)
The Grand Tour, Ron Miller and William K. Hartman (Workman, 1981)

Inside: Seeing Beneath the Surface, Jan Adkins (Walker, 1975)
Underground, David Macaulay (Houghton Mifflin, 1976)
The Whole Cosmos Catalog of Science Activities, Joe Abruscato and Jack Hassard (Scott, Foresman, 1977)

For Young Children

Adventures of Three Colors, Annette Tison and Talus Taylor (Merrill, 1980)
Apples: How They Grow, Bruce Macmillan (Houghton Mifflin, 1979)
The Caterpillar and The Polliwog, Jack Kent (Prentice-Hall, 1982)
Columbus, Ingri and Edgar D'Aulaire (Doubleday, 1955)
Discovering the Mysterious Egret, (grades 1 and up), Jack Denton Scott (Harcourt Brace Jovanovich, 1978)
Guess What?, Roger Bester (Crown, 1980)
How Big Is Big? From Stars to Atoms, A Yardstick for the Universe, (grades K–4), Herman and Nina Schneider (W. R. Scott, 1946)
A Kitten Is Born, Heiderose and Andreas Fischer-Nagel (Putnam, 1983)
Life Story, Virginia Lee Burton (Houghton Mifflin, 1962)
Mousekin's Close Call, Edna Miller (Prentice-Hall, 1980)
Playful Animals (grades 1–6), Dorothy Shuttlesworth (Doubleday, 1980)
Veterinarian, Doctor for Your Pet, Arline Strong (Atheneum, 1977)
Where Do They Go? Insects in Winter (grades K–3), Millicent Selsam (Scholastic, 1982)
Zoobooks, John Wexo (Wildlife Education Ltd., 930 West Washington St., San Diego, CA 92103, 1983)

For Grade-Schoolers

Bet You Can't! Vicki Cobb and Kathy Darling (Lothrop, Lee & Shepard, 1980)
The Chemist Who Lost His Head: The Story of Antoine Lavoisier (Coward, McCann, & Geoghegan, 1982)
Death Is Natural, Laurence Pringle (Four Winds, 1977)
Disease Detectives (grades 4 and up), Melvin Berger (Crowell, 1978)
Einstein Anderson series, Seymour Simon (Viking, a continuing saga)

The Glorious Flight: Across the Channel with Louis Bleriot, Alice and Martin Provensen (Viking, 1983)
The Hidden World: Life Under a Rock, Laurence Pringle (Macmillan, 1977)
How Things Work (National Geographic Society, 1983)
How to Dig a Hole to the Other Side of the World, Faith McNulty (Harper & Row, 1979)
Into Winter, William P. Nestor (Houghton Mifflin, 1982)
Much Ado about Aldo, Johanna Hurwitz (Morrow, 1978)
Mummies Made in Egypt, Aliki (Crowell, 1979)
One Hundred One Questions and Answers about the Universe, Roy Gallant (Macmillan, 1984)
Paddle-to-the-Sea, Holling C. Holling (Houghton Mifflin, 1941)
Paper Airplane Book, Seymour Simon (Viking, 1971)
Robotics, Tony Potter (Hayes Usborne, 1984)
The Science Book (grades 4–7), Sara Stein (Workman, 1979)
See Through the Forest, Millicent Selsam (Harper & Row, 1956)
Shadow Science, Robert Gardner and David Webster (Doubleday, 1976)
Signs and Symbols of the Sun, Elizabeth S. Helfman (Houghton Mifflin, 1974)
Small Worlds Close Up, Lisa Grillone and Joseph Gennaro (Crown, 1978)
The Stars, Irving Adler (Crowell, 1980)
Supersuits, Vicki Cobb (Lippincott, 1975)
To Spoil the Sun, Joyce Rockwood (Holt, Rinehart & Winston, 1976)
Zoos in the Making, Dorothy Shuttlesworth (Dutton, 1977)

For Advanced Readers

The Cay, Theodore Taylor (Doubleday, 1969; Avon, 1977)
The Chemical History of a Candle, Michael Faraday (Viking, 1960). First published in 1861, this was entitled "A Course of Six Lectures on the Chemical History of a Candle."
Dance of the Planets: The Universe of Nicolaus Copernicus (Coward, McCann & Geoghegan, 1979)

Day in the Life of a Marine Biologist (grades 5 and up), William Jasperson (Little, Brown, 1982)
Dinosaurs, Asteroids, and Superstars, (grades 5 and up), Franklyn M. Branley (Crowell, 1982)
Hazel Rye, Bill and Vera Cleaver (Harper & Row, 1983)
Hiroshima No Pika, Toshi Maruki (Lothrop, Lee & Shepard, 1982)
How to Count Like a Martian, Glory St. John (Walck, 1975)
Human Origins, Richard E. Leakey (Lodestar, 1982)
I Hate Mathematics Book! Marilyn Burns (Little, Brown, 1982)
Island of the Blue Dolphins, Scott O'Dell (Houghton Mifflin, 1960)
Journey to the Planets (grades 5 and up), Patricia Lauber (Crown, 1982)
Julie of the Wolves, Jean Craighead George (Harper & Row, 1972)
Kids Whole Future Catalog (grades 5 and up), Paula Taylor (Random House, 1982)
The Little Killers: Fleas, Lice, Mosquitoes, Wyatt Blassingame (Putnam, 1975)
Long Journey From Space, Seymour Simon (Crown, 1982)
Memory: How It Works and How to Improve It, Roy Gallant (Scholastic, 1980)
New Explorers: Women in Antartica, Barbara Land (Dodd Mead, 1981)
The Nuclear Question, Ann E. Weiss (Harcourt Brace Jovanovich, 1981)
Science Experiments You Can Eat, Vicki Cobb (Lippincott, 1972)
Sign of the Beaver, Elizabeth George Speare (Houghton Mifflin, 1983)
Test-Tube Mysteries, Gail K. Haines (Dodd, Mead, 1982)
The Stars: Decoding Their Messages, Irving Adler (Crowell, 1980)
Window into a Nest, Geraldine Flanagan and Sean Morris (Houghton Mifflin, 1975)

ADULT REFERENCES

Books

The Anatomy of Wonder: An Historical Survey and Critical Guide to the Best of Science Fiction, Neil Barron (Bowker, 1981)
Best Books for Children, Pre-School Through the Middle Grades, edited by John T. Gillespie and Christine B. Gilbert (Bowker, 1985)
Best Science Books for Children, compiled and edited by Kathryn Wolff, Joellen M. Fritsche, Elina N. Gross, and Gary T. Todd (A.A.A.S., 1983)
Children's Catalog (H. Wilson, 198–)
The Museum of Science and Industry Basic List of Children's Science Books 1973–1984, compiled by Bernice Richter and Duane Wenzel (Amer. Library Assoc., 1985)
The Cosmic Dancers: Exploring the Physics of Science Fiction, Amit Goswami (Harper & Row, 1983)
Science Books for Children: Selections from BOOKLIST, 1976–1983, ed. Denise Murcko Wilms (American Library Assoc., 1985)
The Science in Science Fiction, ed. Peter Nicholls and David Langford (Knopf, 1983)
Subject Guide to Children's Books in Print (Bowker, 198–)

Periodicals

Appraisal, 3/year, $12 (605 Commonwealth Ave., Boston, MA 02215)
Booklist, 11/year, $40 (American Library Assoc., 50 E. Huron St., Chicago, IL 60611)
Bulletin for the Center of Children's Books, 11/year, $14 (5850 Ellis Ave., Chicago, IL 60637)
Horn Book, 6/year, $21 (Park Square Bldg., 31 St. James Ave., Boston, MA 02116)
Kirkus Reviews, 24/year, $45 (200 Park Ave. S., New York, NY 10003)
The Kobrin Letter: Books about Real People, Places and Things, 10/year, $12 (732 Greer Rd., Palo Alto, CA 94303)
Parent's Choice, 4/year, $10 (Box 185, Waban, ME 02168)

School Library Journal, 10/year, $51 (R. R. Bowker, Subscription Dept., P.O. Box 1426, Riverton, NJ 08077)
Science and Children, 8/year, $28 (National Science Teachers Assoc., 1742 Connecticut Ave., N.W., Washington, DC 20009)
Science Books and Films, 5/year, $17.50 (A.A.A.S., 1776 Massachusetts Ave., N.W., Washington, DC 20036)
Scientific American December issue, $2.50 at newsstands
Selection, 4/year, $12 (Box 5068, Stanford, CA 94305)

Children's Science Book Awards

Children's Book Guild Non-Fiction Award (c/o Washington Post, 1150 15th St., N.W., Washington, DC 20017)
Children's Choices for 198– (International Reading Association, 800 Barksdale Rd., Newark, DE 19711)
New York Academy of Sciences (2 East 63rd St., New York, NY 10021)
Outstanding Science Trade Books for Children (Children's Book Council and National Science Teachers Assoc., 67 Irving Pl., New York, NY 10003)

8 Getting Started with Science Things

You've spent the day working, and you're tired. "Listen, darling," begins your enthusiastic spouse, "I've got a present for you that'll knock your socks off! In this box are directions and all the materials you'll need to learn the ancient art of Samoan basket weaving."

How do you respond? How do you feel?

A child who has never been actively involved in science projects reacts as you do to your spouse's offer: "What are you talking about? A project? I've never expressed an interest in that subject before. How do you know it'll be fun? Have *you* ever done it? It takes energy to learn something new, and I'm tired right now."

Science can be enjoyable, interesting, and satisfying only when one finds an entry point and has the confidence and stamina to proceed.

THE RIGHT SCIENCE FOR THE RIGHT CHILD

There are really two questions involved in judging the quality of a science item: (1) Is the item good in some overall sense? and (2) Will my child enjoy it?

The most obvious ways of judging science things *don't* work. There is no clear relationship between quality and price; consumers are often asked to pay for advertising costs or blister-packed boxes instead of scientific expertise. Place of purchase is also not a perfect cue to quality; stores generally stock what sells or is heavily advertised. Manufacturers assume that the American public is attuned to price-point and eye appeal, not science.

Toy and book producers, who don't know your child, label items by recommended age. This can be terribly confusing, especially for parents of children who are bright and read well. Do not be misled: interest and manual dexterity are not necessarily correlated with IQ. *If you have the choice of purchasing something that is too easy or too difficult, err on the side of "easy."* Success tends to encourage further investigations, while frustration does not. Moreover, doing something faster or earlier does not always mean doing it better, or enjoying it more. In choosing gifts for children, rely more on your observations of the child than on recommended age or grade level.

Furthermore, recognize that there are a variety of pleasures to be found in science toys. The feeling of competence realized in completing a game like RAMI, or an electronics kit, is simply not available to someone studying ant behavior or trying to get crystals to grow. Conversely, there is a challenge and excitement in developing biological or artistic projects—a kind of unpredictability and design—that is not available to the users of the other items.

The best science for children is developed and promoted by people who are aware of the young person's interests, habits, and goals.

Interests

All children have interests that can be informed and broadened through science. It is with these interests that you begin. In practice, it works like this.

Judy is an eleven-year-old who used to spend most of her time thinking about boys, television, or clothes. Last Christmas, her aunt bought her a perfume-making kit, a truly inspired gift. Although Judy still wants to be a movie star when she grows up, the young girl is now often found outside smelling flowers and weeds—everything from camomile to lilac. She is able, with the help of an adult, to distill a number of pleasant fragrances, and has this year begun to sell small bottles of her toilet water to her friends.

To build on Judy's current scientific interests, I recommend a field guide to flowers and weeds, one that discusses the ecology of the plants as well as their identification. I'd help this child notice what grows in cooler, shadier places, and what is common to open fields. Also, because Judy works so comfortably with her distilling device, I'd suggest borrowing a kitchen chemistry book from the library. A flower press might also serve such a child well.

Paul, an eight-year-old, is comfortable with adults, more comfortable perhaps than he is with children his own age. He can do his schoolwork easily and finds himself with hours of unused time, time that he'd like to spend having fun, but how? Paul doesn't like competitive sports, doesn't enjoy large group activities, and isn't spectacularly good with his hands. His mother thought Paul seemed depressed, and went looking for an activity that might fire his imagination. Someone at work told her about the American Radio Relay League, an organization for amateur radio operators. Paul liked the idea of learning Morse code, a kind of secret language; the possibility of communicating with people thousands of miles away was also intriguing. With some coaxing, he wrote to the A.R.R.L., and was sent, a Morse code tape, a detailed information book, and most important, the name of an amateur radio operator in his area who would serve as his mentor and help him get his operator's license. (For more about the American Radio Relay League, see pages 251–252.)

Paul is now an amateur radio enthusiast. He attends local meetings, talks with other people on the air, and is in the process of designing a program on his school's computer to interface with his radio setup. He is also delighted with the fact that were there a local emergency, he would be called upon for help. In short, Paul's hobby has given him pride and expertise.

As you can see from these examples, in helping children locate inviting and appropriate activities, your observations about the child are more reliable than some abstract notion of age or grade level. Think about what this item

Making Science Things Meaningful

As an intelligent and lively adult, you can offer children help in finding and using science things that is not available to even the best shopper. Here's how.

√ Get children to talk to you about what they think is going on as they play with the materials you provide. If you suspect that any of their ideas are amiss, use the research strategies from Chapter 4 to check your and their ideas in an authoritative text.

√ Try to match science books with science things. For instance, Edmund Scientific puts out an excellent plastic model of a wooly mammoth. Generally, children given this item make the beast prance as they provide monster background noises. But if this wooly mammoth is coupled with a copy of Aliki's book *Wild and Wooly Mammoths* (Crowell, 1977), you have an entirely different set of possibilities—the beast can be found by scientists in Siberia, names can be given to each of its parts, or the child can construct a person proportionally small enough to stand next to the wooly mammoth on a shelf. Although Aliki's book is good in its own right, once a reader has an object with which to dramatize and discuss what has been absorbed, the text takes on a whole new dimension. Put simply, the toy and the book add much to each other.

√ The best way to keep science from becoming rote and mechanical is to ask questions about the materials a child is playing with. But adults not schooled in science have difficulty knowing where to begin. Listening to scientists as they approach their work, recurrent queries can be noted: *why, how, how much*

calls upon a child to do and feel, and then ask yourself if the child you have in mind is looking for those particular satisfactions. It is important that early science toys and books create excitement rather than frustration.

Habits Social habits are also important to consider in helping young people choose science activities. Science can be done alone or in a group. There are projects in which the help of an adult is necessary, and there are others children can undertake on their own. Especially if your child is uneasy with science, look for a project with which he/she can work comfortably; if your child doesn't like letting others see his/her mistakes, find something that can be completed alone. If the child enjoys being with others, help locate a science activity that invites shared responsibility and socializing. You might also

and *what would happen if.* Here's how such questions can be applied to materials for young people.

- Generally scientists assume that things in the physical world have a *function.* If you are looking at the wooly mammoth described above, you might ask *why* is it wooly or *what* are the functions of the tusks? "Why" questions needn't be limited to natural objects; in studying a pump, for instance, encourage kids to ask what purpose each part serves.
- Questions about how something works are also relevant to objects of scientific interest. *How* does the wooly covering of a beast provide warmth? What happens if this seemingly useless object on the pump is removed?
- Numerical descriptions of objects are also of interest to scientists. *How old* does the wooly mammoth live to be? How does that compare to the average lifespan of his relative the elephant? Why might that be so? *How long* will it take to pump water out of this basin? How long will it take to siphon out the same amount of water?
- Finally, scientists like to *play with variables* as a way of questioning how things work. In biology this often leads to thinking about genetic mutation and adaptation. In the physical sciences it may well lead to further experimentation: Does it take longer to pump colored water or tap water? Can you use a pump to move sand from one basin to another? Does more baking soda in the vinegar make it fizz longer? Will baking powder work as well? Will lemon juice work as well as vinegar?

consider getting together with another family on a gift. If Beth and Irene are both interested in raising chickens, the girls' parents could chip in on an incubator. Young people who are more comfortable with science are generally willing to vary the circumstances of their play.

Moreover, consider your child's living conditions. If Tekoa has no place to put a geodesic dome model, don't buy one unless you have made arrangements beforehand to donate it to the school, or to construct it as a gift for another child. Similarly, Smog City, California, is not a good place for stargazing. Scientific interest is most easily borne out of the environment in which one lives.

The approach to science things taken by young people who see science as a calling is not exactly that prescribed by the manufacturers. As Daryl, a seventh-grader, explained it, "When I look at a science set I want to know what everything in that box can and can't do." My suspicion, however, is that the aggressive confidence evidenced by this child echoes from years of science play. Just as an avid reader becomes familiar with literary conventions and delights in predicting what will happen next, Daryl, at thirteen, knew how science sets were organized and could ask questions about purpose in a way that someone awed by science and technology could not.

The lesson I take away from Daryl and from children like him is this: If science is revered, and science toys and equipment are proclaimed holy, only the high priests of science are able to experiment and learn. If, on the other hand, science is seen as an everyday event, available to anyone who takes the time and has the energy to play, then science becomes a subject in which anyone can participate. The latter is an important attitude for both science- and non-science-oriented people to develop.

Goals

In a materialistic society such as our own, children asked what science they're interested in generally name an expensive object: a telescope, a microscope, or a computer. As an adult, your job is to shift the conversation from what it is they wish to buy to what it is they wish to do or know. If Effie, who knows nothing at all about astronomy, wants to learn about stars, she is better served by a good book on the subject, a star chart, or a game such as STELLAR 28 or YOTTA than she is by a telescope.

THREE WORRIES—AND HOW TO BE PREPARED

Although you have actively encouraged your children to take on science-related projects, prepare yourselves: the results are not always pleasing. Let's look at three areas in which problems are likely to arise, first, to suggest that such crises are altogether normal, and second, to offer strategies for dealing with these and similar difficulties. An informed parent is less likely to turn temporary distress into long-term aversion.

Messes

Hands-on science tends to be *messy.* Most science projects include lots of pieces—as children collect rocks or shells, muck about with microchips, undertake experiments, or build model rockets, they make a mess.

Suggestions for Coping with Messes

First, talk with the child about the kind of mess he or she is about to make. If there are lots of little pieces that should remain free from dust and out of range of a younger sibling's explorations, find a place where the project can be covered and kept out of harm's way. If the kid is intent on a smelly chemical reaction, do it when company isn't expected, or better yet, some evening when the family is planning to go out to dinner. If the child seems to be building a "dispersal" rather than a collection, find him/her a bookcase with deep shelves to hold a display.

For parents interested in promoting domestic order, Sensaplay makes an extremely useful product—a 23″ square MESSY PLAY AND HOBBY TRAY which has enough uses to make it worth anyone's $7. The plastic board with raised edges can be used for finger painting when the child is young; it's perfect for puzzle pieces; and it serves as a dandy surface upon which to lay out leaf specimens, build a locomotive, or dry homemade paper. The tray slides under a bed for easy storage, and provides a young scientist with a surface which says to everyone in the household: "This is mine—do not touch." The tray can also be covered with plastic wrap if the user is anxious about dust or things tipping out.

The main point, however, is that people who live in families need to understand and respect those objects that other family members see as valuable. Just as you would be furious at a child for trashing a glass figurine that sits on the coffee table "because it's ugly and in the way," a child will be furious with you for discarding what he/she sees as worth saving. A fossil found in the backyard is not just a dirty rock.

For a parent or teacher, no virtue is more important and useful than a sense of humor. Instead of getting angry at the kid who messes up the counter with baking soda and vinegar fizzles, remember that the same experiment will be less interesting at age forty. Moreover, dried fizzles are actually fairly easy to clean up. Breathe deeply at least two times before going on the warpath.

Unrealistic Expectations

Children's expectations for science toys are often unrealistic. Children engaged in reading, writing, or soccer look much like their adult counterparts; there is nothing ostensibly childish about these activities. But when these same young people imagine adult science, they conjure up visions of Cape Canaveral, scientific instruments worth tens of thousands of dollars, and dirt-covered paleontologists victoriously lifting rare dinosaur bones. And in fact, manufacturers who put production dollars into fancy wooden microscope cases and miniaturized, often useless tools are reaping big profits. As children work with these items, however, their hopes for making an easy and important scientific discovery are quickly dashed. (I have often wondered if one of the attractions of the computer is that children and adults, at this historical moment, are working with the same instruments, and the children are often more at home with this new technology than their parents or teachers. In this sense, the technology confirms their adult status.)

Suggestions for Coping with Unrealistic Expectations

Try to prepare a child for what any given science toy or piece of equipment allows him or her to do. Reading the evaluations in this book together should help. Talking to other kids who have the same items might be useful, especially if these friends let the child play with the product you are considering. Libraries may also be of some assistance here. *Consumer Reports* has undertaken evaluations of telescopes and electronics kits, and books such as Aaron Klein's *The Complete Beginner's Guide to Microscopes and Telescopes* (Doubleday, 1980) should also serve you well.

Listen carefully to the potential user of the item. What does she plan to do with it? What does he plan to learn? If you don't know if this item does what these children hope to do, ask the salesperson in a good toy store, museum shop, or science store; the science teacher in your school; or the scientist who lives down the block. The more concrete your questions, the better the answers these people can provide. It is also wise, before purchasing a product, to let the child see what he will be getting. Look closely at the item; talk about what you see. A trip to the store where the item will be purchased often provides the perfect setting in which to elicit from young people what they really hope the item will enable them to do.

In addition, understand yourself the pleasures that are to be had from various science toys. A good chemistry set doesn't help children to blow up the basement, but should acquaint them with chemical principles, for instance, that varying the kind or amount of a substance causes different reactions. This is an interesting idea in its own right. The pleasure to be had in seeing the rings of Saturn is not simply seeing the rings—any good photograph can show you these—but in locating the planet, figuring out which way it is moving in the sky, and in seeing its relation to other heavenly bodies. A telescope is a terrible gift for an impatient kid. Share this information with your children.

Finally, you can build on your child's interest in dramatic science. Gigantic movable parts, loud noises, stink bombs, bubbling and blasting can all lead to more serious interests. Let your child know when the dramatic results hoped for will *not* occur, but when a truly dramatic moment is possible, don't spoil it by predicting the exact result. Later, when a child identifies the conditions that give rise to an experiment, there will be plenty of time to observe and predict together.

Safety

In doing science the issue of safety must be taken seriously. On the one hand, you don't want to paralyze young people with stories of lost limbs and eyes; on the other, you want them to be cognizant of the real dangers inherent in working with specific materials. Generally children do not wish to proceed foolishly, but most assume that claims of danger are exaggerated.

It's difficult for secure, healthy children to believe that anything "bad" could happen to them. Kids are thrilled to test the fates. In the worst of circumstances they use disregard as a way to prove their adult status: "No one tells a grown-up what to do." They take their current well-being as a sign that safety precautions are unnecessary.

In other instances, a child begins working with an item using every safety precaution imaginable, but eventually becomes overconfident and careless. Realizing that the stuff in their chemistry set marked "NaCl" is really table salt, some users feel that they were "tricked" into treating the substance with utmost care. Inappropriately generalizing from this discovery, they then assume that nothing in the chemistry set is worth worrying about. Similarly, they are often unable to differentiate between one situation and another—for instance, taking apart a radio is a fine activity for young tinkerers, but a television can be extremely dangerous. A young person, however, may assume that the two instruments are equally dangerous—or equally safe.

Science toy manufacturers are conscious of the dangers inherent in working with science things, and they specifically warn users about the possibilities of injury. Always assume that manufacturers who put warnings on their products have a reason for doing so. Undoubtedly these warnings should be explained—this, too, is part of science education—but if even no explanation appears, assume that warnings are important, and heed them.

Safety problems generally occur when users don't follow package directions or ignore commonsense precautions. The rocket manufacturers are serious when they tell you to set off rockets in an open field. Sometimes the people who write these precautions are not emphatic enough in stating the importance of safety, and they may be less than skilled at anticipating the assumptions children make (for example, that an electric burner and a gas one are interchangeable). As consumers, look out for this. In the best of all possible worlds manufacturers would provide devices such as safety goggles.

In addition, young people should understand that you encourage their scientific play because you believe they are mature and capable of exercising good judgment. If you find that your child is blatantly disregarding specific precautions or generally not using common sense (lighting an oil lamp with a ponytail cascading about the flame), find a punishment that suits the offense: the child cannot work with these items again until he/she shows greater regard for safety.

Explain that poor safety procedures will not always result in an accident, but that they are taking needless chances. The seat belt example is a good one to use here. Moreover, if they ever have any hope of pursuing science in college or as a career, they will be required to make safety an ongoing concern. Although some accidents are difficult to avoid, most are the result of careless practice.

If you know that your child is a bit absentminded, work with him or her until you feel confident that the recommended safety precautions have become virtually automatic. Even if your child is extremely reliable, do not allow him or her to work with volatile substances or in dangerous places unless you are within earshot.

If you depart from recommended safety practices yourself, explain to the child how and why you have made your decision to do so. Tell the child that this is not a decision you feel that he or she is ready to make. In addition, try to articulate those safety precautions that you take. For example, I unplug the iron, even after it's been turned off, and leave it on a high shelf; I do not want it within reach of the baby. Point out similar precautions that you observe.

When you visit a scientific or technological site, note how industrial safety concerns are manifested. Chemistry labs have hoods, standardized ways of marking toxic substances, and methods of disposing of chemical waste. On a construction site, workers wear hard hats and use scaffolding and belted devices for multi-story climbing. Geologists, biologists, electronics people, and physicists all maintain their own safety standards.

With all this talk about avoiding disaster, mention should be made of another group with safety-related problems. There are children who are so afraid of electricity, and chemicals, and saws, and flashing buttons that they are reluctant to work with any hands-on materials. Do not deny their concerns, but rather commend these kids for their foresight. Go on to tell them that they have a real advantage in working with potentially dangerous materials. If their fears are deep-seated and debilitating, you might steer them toward science-related activities in which issues of safety are minimized—

computer work, for instance, or model construction. Recognize that these children are generally fearful because they have taken on adult fears; try working together on an electronics project, and you will both be amazed by your success. If you find yourself completely unable to cope with the machine age, find a third party who can work with you both.

Finally, talk to the child about those particular substances and experiments he or she is afraid of. Build on strengths, don't push too hard, and don't tease a child with serious fears. A kid who is afraid to light a match is more likely to start a fire than one who has had experience and can think rationally about what to do in case of an emergency. Let the child see that you are there to help, that you are his/her advocate, and that together, eventually, he/she will be able to do all those neat projects you can both imagine.

I have two more comments on safety, and these are difficult to deal with. There are, in fact, a few books and materials that recommend unsafe experiments. It is not so much that the experiments themselves are completely unsafe, but rather that the chance for error is high and the results of a misunderstanding are extremely serious. For instance, A. Harris Stone in his well-regarded book *Science Project Puzzlers: Starter Ideas for the Curious* (Prentice-Hall, 1969; recently reissued in paperback), recommends that children undertake an experiment with a fluorescent light bulb. Although the experimenter is warned that if the bulb breaks it will implode, and that the white coating inside is dangerous, still, accidents can happen. On page 34 of this same text children are directed to "heat a bowl of carbon tetrachloride." Even though they are told to heat it gently over an electrical element and not to breathe the fumes, the possibility of error and the magnitude of the danger inherent in a mistake are indeed distressing. For me, the skull and crossbones danger symbol and the note that these experiments should only be "undertaken with the supervision of a knowledgeable adult" are inadequate; my chemical experts say no one outside a lab should heat carbon tet.

Unfortunately, there is no sure way for a novice to catch all these possibilities. If you are particularly concerned about issues of safety, I recommend the following:

- Ask the bookseller or librarian if they have kept up with reviews on this particular text. Have there been any comments about safety? The American Association for the Advancement of Science in their reviewing journal *Science Books and Films,* and *Appraisal,* a reviewing source that uses scientists as well as librarians to evaluate materials, are particularly informative in this regard.
- Read over the book in question. If your common sense suggests that something is amiss, check it with a local science teacher or have your librarian check with an expert.
- Work with your child. Even if you are not sitting down together to work on a given experiment, make sure that you are nearby in case of an emergency. If the child is off on a jaunt collecting rocks or tracking birds, encourage him or her to go with a friend and check with each other at regular intervals.
- Stress the importance of following recommended procedures. If directions say to heat something on an electric stove, *do not* assume that a gas stove will work just as well. If a book says that two mushrooms are easily confused, don't take a bite and hope for the best. These warnings are serious.
- As a general rule (but unfortunately not infallibly), books and sets with very recent copyrights tend to be more safety-conscious than materials from twenty to thirty years ago.

The other taxing problem with safety is that some of the most dramatic and exciting experiments simply are not safe. In many instances you trade glamour for safety, and this is a difficult trade for young people to make willingly. There are no easy solutions. Certainly you can explain the problem to children, and some will understand. You can also try to interest them in safer aspects of science play: "Chemistry sets will not allow you to blow up the kitchen, but they will let you see the great importance of what you add, and how much you add."

Another alternative is to let a child read about, and thus vicariously enjoy, the dramatic and unsafe. Science fiction and biography can be useful here.

Finally, realize that certain toys or pieces of equipment lead to more overtly thrilling results. If your child is out for excitement, work with him or her in learning the safety precautions the manufacturers of Estes rockets, for example, suggest. I stand firmly by the rule that the more dangerous the toy, experiment, or outing, the more cautious and deliberate the child needs to be.

As I said before, hopeful, healthy children generally don't believe that anything will really harm them—especially things that look ordinary. On a recent tour of a molecular biology lab, I was shown a gel which supposedly held some cloned cells. The substance looked as if it were made of clear gelatin or transparent rubber, and I went to touch it, simply to see which it was more like. "You'd better not," said the doctoral student. The gel, it turns out, adversely effects the nervous system . . . and yet it looked so harmless.

Every day each of us makes decisions about which dangers are worthwhile and which are not. People live in areas where the cancer rate is much higher than the national average because they can find suitable employment there; they eat bacon and hot dogs although they know that these products contain nitrites; they drive cars although they know about the dangers inherent in automobile travel. In evaluating dangers, we ask ourselves three questions: (1) Are the dangers we court worth the risk? (2) Have we done everything possible to minimize the risk? and (3) Do the risks we take unwittingly expose others to unsafe conditions—that is, are we being responsible?

These same three questions should guide your decisions about science play. Although coating a coin with mercury may be fun, the thrill of the shine is *definitely* not worth the danger of messing with a dangerous chemical. On the other hand, I know that rockets can be dangerous, but I believe that the value an interested and reliable kid can realize from an intense interest in model rocketry is worth the dangers inherent in the activity. In a world where technological solutions are sought, we need to teach our children about the pleasures and the dangers that science and technology may bring. There is no better place to begin this education than at home.

Tips for Encouraging Scientific Play

√ Notice what makes it difficult for a child to begin science play. If the TV is a problem, set it up in a room where it is harder to get to and less comfortable to watch. If putting materials in order is difficult, a parent can help gather and organize necessary equipment.

√ Set aside a specific time for playing with those items kids "wish they had more time for." Remind little Dennis how much he enjoyed his last effort.

√ Be conscious of annoyances. Keep the baby from crawling across the table where eight-year-old Katie is working. If Miguel needs quiet, find a room where he can shut the door, or buy him earplugs or headphones.

√ Have specific goals in mind. "When you finish your model of the heart, let's take it to the science center downtown and compare it to their display." "Those wildflowers you're pressing would look spectacular on greeting cards. Aunt Ruth's birthday is coming up next month and you were looking for a gift idea."

√ Work with rather than against children's interests and abilities. If Jeremy likes to eat, experiment with baking chemistry. If Dora is able to concentrate on two things at a time, take out a building set as she watches her regular television program.

√ Ask for children's input. If you wonder whether Oliver would like the microscope you are considering, ask him. If you want to find an enjoyable, science-oriented activity for the entire family, ask the children to help plan it.

√ Develop shared scientific interests. If you've ever thought that it would be fun to build your own stereo or TV, consider learning electronics with your child. If the family has purchased a computer, work on programming with your son or daughter; equal footing serves both of you well.

√ Engage children in meaningful, adult conversations. Many young people complain about grown-ups who only ask questions that require demeaning, parrot-like responses. In science especially, it is too easy, and too demoralizing, to set up one party as the expert and the other as the information sponge. Vary your approach.

√ Avoid embarrassing comments. Children who are having a hard time taking themselves seriously certainly do not want you to tell "cute" stories about their misconceptions or mistakes. If they see you as trustworthy, sharing increases.

√ Be forgiving. If a child mistakenly insists upon, buys, and is bored by an item, don't exaggerate or prolong the suffering. Just because she doesn't like a poorly designed physics set doesn't mean she won't enjoy the study of physics.

IN SUM . . .

Most parents, teachers, librarians, and youth group leaders feel they are ill-equipped to help children with scientific and technological projects. I hope that this chapter has made you less afraid. Even the most scientifically naive adult who is willing to read and experiment can expose children to and help them enjoy science. Turn your lack of experience into an asset, and learn with your child. In thousands of ways we ask children to take intellectual risks, to trust their teachers, to expose themselves to criticism, to make mistakes. This is your moment to suggest that those risks are worthwhile, and to show young people that learning is a lifelong process.

9 Where to Buy What

The first part of this book suggests a variety of things you can do to promote an interest in science and technology—places to visit, ideas to discuss, research projects to pursue. The remainder of the book is devoted to an evaluative listing of things you can buy which encourage children in thoughtful, scientific play—toys, books, posters, and magazines.

TOY SUPERMARKETS

Americans spend most of their shopping hours in supermarkets of one kind or another. These outlets for groceries, drugs, toys, and books are designed to give consumers variety, convenience, and reduced prices. Companies whose names one recognizes from television and magazine advertisements, for instance Skilcraft, Fisher-Price, Ideal, Lego, or Golden Books, can be found in great profusion at toy supermarkets. Supermarkets are *not* likely to carry books referred to here as "trade" titles—those generally found in libraries or bookstores—or toys that have not been promoted through media campaigns.

SPECIALTY SHOPS

In large and medium-sized cities one finds assorted specialty shops—children's bookstores, toy stores, science specialty outlets, and electronic parts companies. Small toy shops are likely to carry products produced by small companies, and children's bookstores keep their shelves well stocked with trade titles, many of which are now available in paperback editions. Knowledgeable and interested merchants create stores that are worth frequenting for at least three reasons:

- Specialty shops often stock merchandise not distributed through mass-market chains. They are prepared to deal with small companies, imported products, and special requests.
- Specialty-shop owners are often able to answer questions about what's inside the blister-packed box: For whom is it intended? What does the child get to do? Who recommends it? Have there been any complaints from other consumers?
- Because they are small, specialty shops must maintain the goodwill of customers. Frequently they will special-order merchandise. Often their salespeople are able to figure out why something isn't working. In time, they can learn about your family's tastes and make appropriate gift suggestions.

Even though many such establishments occupy small buildings and their overall inventory is less than that of toy and book supermarkets, for quality books and toys, telescopes, microscopes, and binoculars, they are, in fact, better supplied. Although you may (or may not) pay more at a small retail establishment, remember that in shopping "small" you receive an important service as well as a product.

To determine if a bookstore is stocking quality informational literature, use these series as checkpoints: Little, Brown's Paper Books, Crowell's "Let's Read and Find Out" science books, and the EDC Publishing (Usborne Hayes) science series.

To determine if a science store is stocking a full line of science toys, check for items from Suitcase Science or Battat, Learning Things, Thomas Salter/NSI, and Edmund.

Science catalog houses such as Edmund Scientific, The Nature Company, and Energy Sciences have retail outlets. In addition, there are shops devoted completely to science, for example Science Hobbies in Charlotte, NC, Science Things in Farmington, MI, or the Educational Modules shop at The Rochester Museum of Science and Industry. If you are fortunate enough to live near or visit one of these places, go in and browse.

MUSEUM SHOPS

Still another excellent source for purchasing science toys and books is the science museum shop. These stores cater to two groups: people looking for an inexpensive souvenir, and people in search of an intelligent and thoughtful birthday or holiday gift. I like buying things at museum shops because my dollars help support the museum. A progressive museum shop may also help patrons learn about and use items through museum programs.

If an item you purchase at a museum seems less than adequate, you can go back to a knowledgeable person for advice. Remember, however, that museum stores exist to make a profit as well as to educate the public. If there is an ongoing demand for beautifully packaged, overpowered microscopes (see Chapter 12), for instance, they may carry these relatively worthless items.

A good museum shop should have an excellent selection of children's science books. If yours doesn't, speak to the store manager. Tell her/him about the reviewing sources mentioned here, and ask for help in making materials available.

GOVERNMENT SOURCES

Agencies of the federal government carry items of interest to science enthusiasts. The National Aeronautic and Space Administration (NASA) from time to time sells or gives away space things; recent items included seeds from space and a model launching pad.

The Government Printing Office (GPO) has more than 17,000 books currently available; approximately 2000 titles are added annually. Some of these books are intended for children. Sample titles include: *My Wetlands Coloring Book* (1980) for children in grades K to 3; *Meeting with the Universe: Science Discoveries from the Space Program* (1981), for children in grades 5 and up, by Bevan M. French and Stephen P. Maran; and Jean Craighead George's *Everglades Wildguide* (1972), also for advanced readers. GPO bookstores are located in Washington, DC; Atlanta; Birmingham; Boston; Chicago; Cleveland; Columbus; Dallas; Denver; Detroit; Houston; Jacksonville; Kansas City, Missouri; Los Angeles; Milwaukee; New York; Philadelphia; Pueblo; San Francisco; and Seattle. For a catalog of current publications, or to order books directly, write: Superintendent of Documents, Government Printing Office, Washington, DC 20402.

National Laboratories that house museums (such as Brookhaven National Lab in Upton, NY, the Oak Ridge Lab in Tennessee, or the Lawrence Livermore Lab in the San Francisco area) often have the best prices around for the items they carry. These are great spots for holiday purchases.

CATALOG SHOPPING

The convenience and variety available from catalogs is hard to replicate in any town or city. The following catalogs carry a wealth of science items. Specialized catalogs, such as those devoted to electronics or computers, are listed in this book under the subject chapters. Specialized magazines, for instance *Sky & Telescope* and *Earth Science* (again, see subject listings), also carry numerous advertisements for hard-to-find items. *Note:* Catalog prices vary significantly. Moreover, a price may be high on one item and low on another. Since most companies happily send catalogs for free, shop and compare.

General Toy Catalogs

The following catalogs carry a good selection of science items as well as the usual dolls and teddy bears.

Childcraft (20 Kilmer Rd. Edison, NJ 08818; toll free 800-631-5657). This general toy catalog carries a good four-page section on quality science items.

Just for Kids (Winterbrook Way, Meredith, NH 03253; 603-279-7011). Among the dolls, stuffed animals, and preschool items you'll find a nice collection of science things.

Play 'n Peace (P.O. Box 775, Medford, NJ 08055). This $1 catalog includes a variety of educational toys. Minimum order $15.

Toys to Grow On (P.O. Box 17, Long Beach, CA 90801; 213-603-8890). The educational consultant for this general toy catalog, Marilyn Sloan, Ph.D., does an excellent job of finding and putting together a fine collection of science toys. Items marked "Lakeshore," the school division of Toys to Grow On, can also be purchased here. Catalog free.

Science Catalogs for the Home Consumer

Aquarium and Science Supply Co. (101 Old York Rd., Jenkintown, PA 19046; toll free 800-453-3333). Birdhouses, small-animal cages, and lots more at 35% to 50% below standard biological supply house prices. Catalog free.

Edmund Scientific (101 East Gloucester Pike, Barrington, NJ 08007; toll free 800-257-6173). The best-known source of home science gifts, Edmund sends their extensive and useful catalog free upon request. Not ony does Edmund carry their own products, they also have an excellent selection of science toys and equipment, including some salvage and industrial materials. Store managers are knowledgeable and willing to answer consumer's queries.

Energy Sciences (The "Wonderbook," 16728 Oakmont Ave., Gaithersburg, MD 20877). This $2 catalog includes items related to understanding, appreciating, and saving energy.

Insect Lore Products (P.O. Box 1535, Shafter, CA 93263). A small but very useful catalog of nature-related items.

Nature Company (P.O. Box 7137, Berkeley, CA 94707; toll free 800-227-1114). This attractive catalog highlights items designed to help people enjoy and celebrate, not destroy, nature. These include an elegant selection of posters and T-shirts, binoculars, toys, and books. The several stores of the same name, located up and down the California coast, include an even wider selection of gift items and juvenile trade books on natural history.

Radio Shack (Tandy Corp., 1400 One Tandy Center, Fort Worth, TX 76102). The Radio Shack catalog includes their own wide selection of electronics, physics, and computer toys. Catalog free from local outlets or by writing direct.

Museum Store Catalogs

Discovery Corner (Lawrence Hall of Science, University of California, Berkeley, CA 94720). Advertised items change regularly; catalog free upon request.

Discovery Shop (American Museum of Science and Energy, 300 S. Tulane Ave., Oak Ridge, TN 37830). Energy-related books and toys. Catalog free upon request.

Exploratorium Catalog (Exploratorium Store Catalog, 3601 Lyon St., San Francisco, CA 94123). At holiday time the Exploratorium puts several toys together in kits with explanations. They also carry a good line of items of general scientific interest. Catalog free upon request.

Museum of Science, Boston (Museum Store, Science Park, Boston MA 02114). A foldout sheet of gift ideas, sent free upon request.

Catalogs Catering to the School *and* Home Market

Hubbard Scientific (P.O. Box 104, Northbrook, IL 60062; toll free 800-323-8368). Although Hubbard makes their catalog available only to schools, they are willing to fill orders for items listed in this book. A reliable source for an interesting collection of items.

Jerryco (601 Linden Pl., Evanston, IL 60202; 312-475-8440). An interesting collection of surplus items, and a great read. In the last issue, there were what looked like a good mineral collection, inexpensive solar cells, and several funky motors. The Jerryco catalog exercises the imagination, and provides inexpensive materials for the science hobbyist. Catalog "half a buck."

Learning Things, Inc. (68A Broadway, P.O. Box 436, Arlington, MA 02174). Magnifiers, microscopes, cardboard carpentry tools and accessories, sold direct. Free catalog; minimum order $10 before shipping and handling.

Merrell Scientific (1665 Buffalo Rd., Rochester, NY 14624). Catalog features a wide selection of chemistry, biology, and hobby supplies, as well as rockets; can be purchased for $2 from Merrell suppliers or directly from the company. Minimum direct order: $20.

Nasco (901 Janesville Ave., Fort Atkinson, WI 53538; toll free 800-558-9595). This free catalog provides an excellent source of science things, including plants, toys, and school supplies.

Schoolmasters (745 State Circle, Box 1941, Ann Arbor, MI 48106; toll free 800-521-2832). Catering largely to a school audience, this catalog carries many products of interest to parents. Ask for the free science catalog. Service charge on orders: $3.

Tops (10978 S. Mulino Rd., Canby, OR 97013). This series of hands-on activity guides, designed primarily for classroom use, is based on readily available materials and works well at home. Catalog free.

Transtech (Creative Learning Systems International, 9889 Hibert, Suite E, San Diego, CA 92131). This free catalog includes a four-star collection of science books and toys, including FISCHERTECHNIK, LEGO TECHNIC, and computer software.

Catalogs for Institutional Use

Carolina Biological Supply Company (2700 York Rd., Burlington, NC 27215; 919-584-0381). The 976-page Carolina catalog, which includes books, games, specimens, lab equipment, and lots more, is distributed free to schools but must be purchased for about $8 by individuals. Elementary school science materials from Carolina Biological are previewed in a catalog distributed free while supplies last.

Fisher Scientific (4901 W. LeMoyne St., Chicago, IL 60651; toll free 800-621-4769)

Ideal (11000 S. Lavergne Ave., Oak Lawn, IL 60453)

Sargent Welch (7300 N. Linder, Skokie, IL 60077; 312-677-0600)

Wards Natural Science Establishment, Inc. (5100 W. Henrietta Rd., P.O. Box 92912, Rochester, NY 14692). This full-line science catalog is especially well liked by geologists.

For an extensive list of school-oriented science catalogs see the annual January issue of *Science Teacher* or *Science and Children,* two publications of the National Science Teachers Association.

10 General Science: A Pinch of This, a Handful of That

Fig. 1

Most kids don't think of science as biology, chemistry, physics, or engineering. Instead, they hold some generalized notions of what scientists do—scientists experiment, they are concerned with objectivity, they observe and classify and speculate about why certain things occur. I have used traditional academic divisions in this catalog to classify recommended toys and books because they are useful to the adults looking for items. Certain items, however, cannot be classified as anything but general science.

Fig. 1. Among the best places to explore "general science" are participatory science and technology centers. This photo from San Francisco's Exploratorium shows families playing with the Catenary Arch exhibit as they learn about the strength of the natural shape taken by a hanging chain.

HANDS-ON

Measuring Tools

For All Ages

Thermometers, scales, measuring cups, rulers, barometers, balances, a stopwatch, and other equipment that provides quantitative descriptions of events make wonderful gifts for families interested in estimating, predicting, and recording. Located in hardware stores or through science supply houses, these items should be sturdy enough for use by children as well as adults.

For Young Children

Fig. 2

Fig. 2. Calipers (about $9 from Hubbard) are used to determine thickness, diameter, and distance between surfaces. For children who enjoy measuring, they provide a nice change from the standard ruler and string method.

Fig. 3. The PRIMER BALANCE, available from Nasco for $20, is a precise, durable, and sensitive instrument which enables young children to learn about balance, compare mass and weight, and measure liquids as well as solids. Objects can be placed anywhere in or on the plastic buckets without affecting accuracy.

Fig. 4. An abacus invites children to count, arrange numbers in groups, add, multiply, subtract, and divide. It also provides a visceral sense of quantity without lots of little pieces to lose around the house. This colorful version sells for $20 from Toys To Grow On / Lakeshore.

For people who *can* keep track of little pieces, few educational toys are better than CUIZINAIRE RODS, the small, proportionally shaped, colorful wooden rods available in school supply catalogs or better toy stores for approximately $57.

For Older Children

Fig. 5. Balances are the measuring devices most often used by scientists. An accurate plastic model, available from Hubbard Scientific for $10, is a comparative bargain.

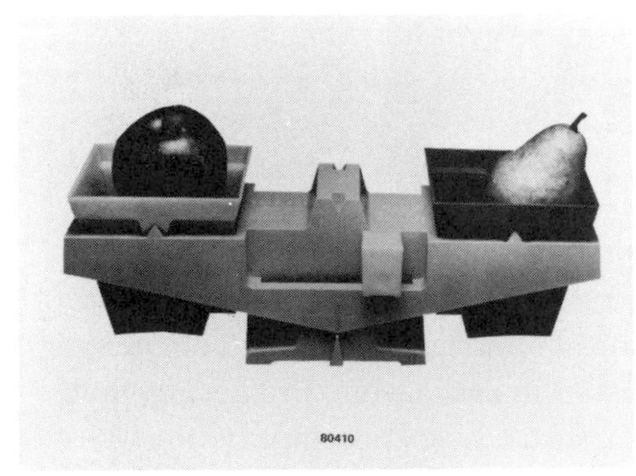

Fig. 3

Fig. 4

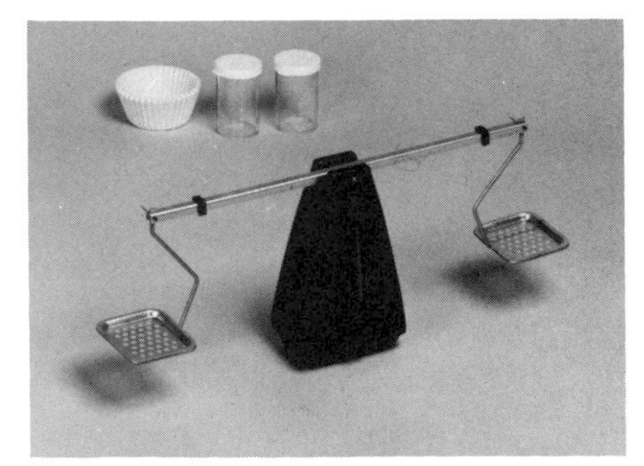
Fig. 5

Toys for Pretending

Animal puppets, STAR WARS figures, toy robots, toy physician's kits, and dolls invite children to try out roles and have conversations about matters of scientific import. Like all authors, children look for content to make their stories more realistic, frightening, reassuring, or humorous. "Pretends" give parents a spectacular opportunity to engage children in conversations about social responsibility ("If we bring back all this gold from planet Yarsow, we'll be rich. But there are creatures here who eat nothing but gold. What should we do?"), as well as science content ("Let's go spelunking. What do we need to take? How can we find our way out of the cave if we get lost?"). Most children are helped in such fantasies by props—costumes, implements, and sets. In this sense, such accoutrements are hands-on toys.

Learning magicians' tricks can also help children understand illusion, an important concept in science. Nothing real, in fact, defies the laws of nature; only those who lack information or explanation can be fooled.

General Science "Labs"

Both Gabriel and Skilcraft offer general science kits which include a smattering of experiments in chemistry, geology, and biology. In some ways these are the most interesting kits on the market. When a child tires of mixing chemicals, he/she can start fresh testing minerals for hardness, or experimenting with hydroponics. There are, however, problems with both sets. Apparently the included experiments were simply borrowed from individual geology, biology, and chemistry kits, and little attempt has been made to show interrelationships between disciplines. For instance, Gabriel's TRI-LAB PAK discusses oxidation and reduction, an important topic in chemistry, only in its mineralogy section.

The Gabriel TRI-LAB PAK ($14) contains a microscope, with suggestions of several objects to view, as well as an extensive section on "chemical magic" and the chemistry of acids, bases, and salts. You are expected to add 5% hydrochloric acid (muriatic acid), denatured alcohol for an alcohol lamp, and blue (cobalt) glass. The best experiments in the kit are found in the geology section: ten mystery rocks and minerals are identified through chemical and

Fig. 6

physical tests, while still other experiments help verify those identifications. *Note:* Add an eyedropper to this kit.

Fig. 6. Skilcraft's QUADLAB ($22), shown here, includes a frog in formaldehyde, dissecting knife, and brine shrimp eggs as well as the microscope. The diagram for identifying frog parts would best be replaced by a book on dissection. Also, read through the hydroponics and brine shrimp experiments before beginning, since yeast will be needed at some point. The chemistry section includes "chemical magic" as well as better-explained sections on solutions and colloids; acids, bases, and salts; and exchange reactions. The Jolly balance used to measure the specific gravity of rocks adds a nice quantitative aspect to this kit. *Note:* Both sides of the measuring spoon in this kit can be used; use the smaller side to avoid spilling dangerous substances.

Both sets are also weak on equipment, a situation which can be remedied with a little know-how:

- Replace inadequate microscopes with a good pocket magnifier (see pages 153–154).
- Buy a test tube brush, needed for cleaning glassware and scraping out chemicals, since a brush is not included in either set. Take one of the tubes along as you look for a tiny (doll's?) bottle brush.
- Buy corks or stoppers since both sets lack these. A child will automatically put a finger over the tube when shaking a solution, which is bad practice and can be unsafe. A cork is also useful if one wishes to save a solution.

Monthly Science Gifts

Things of Science is an experiment-of-the-month club for ten- to sixteen-year-olds, providing materials, an explanation booklet, and directions for demonstrations and experiments. The kits change from year to year, although experiments run three to four years earlier are occasionally revised and repeated. Some kits are designed to develop an interest in a subject, such as "shells," while others, such as "levers and pulleys," are project-oriented. Other units have focused on sound, aerodynamics, gravity, mathematical paper folding, and optical illusions. The variety of experiments, the quality of the explanations, and price of this program make it an excellent choice. For a one-year subscription (approximately $17.50 for the twelve "things"), write Things of Science, RD #1, Box 130, Newtown, PA 18940.

IN PRINT

Experimental Miscellany

The SCIENCE PARTICIPOSTER is a 24″ × 36″ oaktag wall hanging that includes crossword games, information questions, observation and measurement challenges, and more. The easiest questions can be answered by a lively first-grader, while the more difficult items are suitable for a good middle-school student. Cost is $4, from Ampersand Press, 2603 Grove St., Oakland, CA 94612.

A boxed set of 135 projects for children age seven and above, called THE ELEMENTARY SCIENCE ACTIVITY PACK (from Educational Insights of Compton, CA), features experiments in magnetism, electricity, air, heat, sound, light, weather, water, simple machines, and chemistry. Available through Nasco, Toys To Grow On, and other science supply houses for about $10. Directions are clear, and the card format is handy. Good experiments and explanations at a reasonable price.

Fig. 7

Fig. 7. A wall-hanging science calendar designed by the Smithsonian Institution's Family Learning Project provides an experiment-of-the-month. Ordinary materials are used to make a sundial, "pizza map," ginger ale, an ant farm, and more. Directions are clear, illustrations inviting, and the science ideas are interesting, unthreatening, and fun; $9.95 per copy (plus $2.50 shipping and handling) from GMG Publishing, 25 West 43rd St., New York, NY 10036.

Note: For 2 to 6 copies the price stays the same, but the handling charge drops to $1; consider combining your order with that of a friend. Significant multiple-copy discounts are available for schools, PTAs and other qualified groups.

Magazines

For All Ages

Magazines and books intended primarily for adults—*Science 86* or *National Geographic,* for instance—which include attractive, clear illustrations may interest even young children. Having publications such as these about the house also sends out a clear message to young people: Science is interesting and accessible. A similar function is served by coffee-table editions of books on the TV series *Nova,* Jacob Bronowski's *Ascent of Man* (Little, Brown, 1974), Carl Sagan's *Cosmos* (Random House, 1980) or Jon Darius's *Beyond Vision* (Oxford, 1984), a fascinating collection of photos from the inaugural exhibit at the new Science Museum in London.

For Young Children

Scienceland—8/year, $12 (501 Fifth Ave., Suite 2102, New York, NY 10017). The one, true, general science magazine for preschool to third-grade children. Designed to "nurture scientific thinking," the periodical is appealing, highly pictorial, and gives children practice in observing, classifying, reasoning, and enjoying science.

Other general magazines for young children, *Highlights,* for instance, also include informational stories. Not to be missed are the several excellent nature magazines for small children listed on page 146 of the biology section.

For Grade-Schoolers

Cricket, 12/year, $17.50 (Box 2670, Boulder, CO 80322). Designed for children six to twelve, this high-quality publication includes some science as well as a good deal of literature and history. *Cricket* can be seen as the children's counterpart to *The Atlantic*—it is thoroughly undidactic, literate, and broad-based in its approach.

National Geographic World, 12/year, $9 (Department 00480, 17th and "M" Streets, N.W., Washington, DC 20036). National Geographic's factual magazine for children ages eight to thirteen focuses on "the things kids like most: wild animals, pets, hobbies, sports, and, of course, other kids." Full-color photos, lots of activities, and well-written copy make this one an excellent choice.

Owl, 10/year, $14 in U.S., $12 in Canada (Young Naturalist Foundation, 59 Front St., Toronto, Ontario M5E 1B3, Canada). This highly recommended "participation" magazine for ages eight to twelve presents articles on nature as well as features on technology and the history of science. It is attractive, well-written, and humorous (the animal centerfold is a monthly favorite).

3-2-1 Contact, 10/year, $11 (E=MC Square, P.O. Box 2932, Boulder, CO 80321). A "people-oriented" science magazine for kids age eight to fourteen, *3-2-1 Contact* closely follows the public television show on which it is based.

For Middle-Schoolers

Penny Power, 6/year, $9 (P.O. Box 1906, Marion, OH 43302). Published by Consumers Union, this magazine for nine- to thirteen-year-olds reports on items children use or buy—everything from breakfast cereals to bicycles. In May 1984, an article on science toys appeared.

Science Challenge, 9/year, $9 (Curriculum Innovations, Inc., 3500 Western Ave., Highland Park, IL 60035). Intended primarily for seventh- to twelfth-grade classrooms, the issues discussed in this periodical are designed to address the basic questions of "scientific literacy"—What's new? What social changes are signaled by scientific change? What, if anything, do we want to do with this information?

Science World Magazine, 18/year, $6 (902 Sylvan Ave., Englewood Cliffs, NJ 07632). This classroom magazine for grades 7 to 10 is designed to "help develop thinking and research skills as well as science vocabulary." It includes current information as well as advertisements and an entertainment section.

Parents, teachers, and librarians often have difficulty locating informational literature for small children. In fact, many titles which could be called "science" are shelved among the picture books, while others that are easy enough for the very young are found in the "science" sections. The best general science books for young children encourage them to look again, and this time more carefully, at the world in which they live. Here is a list of books which are sure to help in this regard.

For Young Children

Barton, Byron, *Airport* (Crowell, 1982).
Like all of Barton's books, this has energy, simplicity, and uncluttered information to recommend it.

Dabcovich, Lydia, *Follow the River* (Dutton, 1980).
Water flows downstream past cities and countryside with poetic clarity.

Gibbons, Gail, *New Road, Clocks and How They Go* (Crowell, 1979), *Fill It Up! All About Service Stations* (Crowell, 1985) and (written by Joanna Cole and illustrated by Gail Gibbons) *Cars and How They Go* (Crowell, 1983).
These bold and colorful books make everyday objects and events dramatic and comprehensible.

Hoban, Tana, *Is It Rough? Is It Smooth? Is It Shiny?* (Greenwillow, 1984).
A wordless picture book of real things with startling textures—elephant hide, shiny pennies, cotton in the boll, and lots more.

Huinberstone, Eliot, *Finding Out About Everyday Things* (EDC, 1981).
Provides cogent explanations about *How Things Go, Things Outdoors,* and *Things at Home.*

Rahn, Joan, *Holes* (Houghton Mifflin, 1984).
A conceptually sophisticated look at common spaces, this one invites lively conversations about classification and definition.

Rockwell, Anne, *Toolbox* (Macmillan, 1974).
Clear, bold, informative, and utterly appealing.

Schwartz, David M., *How Much Is a Million?* (Lothrop, 1985).
The need for recognizable mathematical terms, (a fish bowl for a million goldfish "would be large enough to hold a whale") is truly satisfied in this book for early elementary schoolers.

Selberg, Ingred, *Nature's Hidden World* (Putnam, 1984).

Pull-tabs and pop-ups on six woodland scenes reveal insects and small beasts, hidden as they would be outdoors.

Testa, Fulvio, *If You Look Around You* (Dial/Dutton, 1984).
The precision of language—point, line, circle, sphere—is celebrated in this whimsical book for very young children.

For Grade-Schoolers

This list should prove beyond a doubt that books can compete favorably with toys.

Allison, Linda, *The Wild Inside* (Scribner, 1979).
At-home activities teach kids about physics, geology, weather, electricity, and natural history.

Herbert, Don, *Mr. Wizard's Supermarket Science* (Random House, 1980).
Glue from milk, vinegar rocket launchers, salad-oil soap, and more make the connections between ordinary substances and laboratory science more evident.

Robbins, Pat, *Far-Out Facts* and *More Far-Out Facts* (National Geographic, 1980 and 1982).
Two fascinating volumes for lovers of "gee-whiz" information.

Schneider, Herman and Nina, *Science Fun for You in a Minute or Two: Quick Science Experiments You Can Do* (McGraw-Hill, 1975).
These twenty-six physical science experiments are easy and satisfying—an excellent alternative to the "I can't think of anything else to do" TV show.

Stein, Sara, *The Science Book* (Workman, 1979).
Includes a variety of physics, chemistry, and biology experiments using objects of general interest: your own body, animals, pets, food, and junk.

Woods, Geraldine and Harold, *The Book of the Unknown* (Random House, 1982).
If your child likes speculation, especially about questions to which there are no known answers ("Can I live to be a hundred and fifty?"), this one presents interesting-to-discuss working theories.

For Middle-Schoolers

These books, intended for advanced readers, are excellent. With some help, they are also appealing to children who read less well.

Cobb, Vicki, and Kathy Darling, *Bet You Can't! Science Impossibilities to Fool You* (Lothrop, Lee and Shepard, 1980).
Cobb and Darling give the setup times, clear directions, equipment lists for and adequate explanations of sixty science tricks.

Haines, Gail K., *Test-Tube Mysteries* (Dodd, Mead, 1982).
Scientists ferret out what we call "answers"; the passion, wrong turns, and fortuitous accidents make for a great read.

Janeczko, Paul B., *Loads of Codes and Secret Ciphers* (Macmillan, 1984).
Spies, hoboes, cowboys, Indians all used secret codes. Systems are explained, and kids are challenged to figure out, vary, and create their own messages.

Lambert, Mark, et al., *All Colour Book of Science Facts: Our Earth—Plants and Animals—The Human Body—Science and Technology* (Arco, 1980).
Information-hounds age ten and older will have trouble putting this one down.

Renner, Al G., *How to Build a Better Mousetrap Car—and Other Experimental Science Fun* (Dodd, Mead, 1977).
For kids who like the engineering activities from "Olympics of the Mind"—hands-on puzzlers to test your design skills—this one is a must.

Schneider, Herman, *How Scientists Find Out about Matter, Time, Space, Energy* (McGraw-Hill, 1976).
One of the few books that accurately describes the scientific method. Clear, interesting, unpretentious.

Schrier, Eric W., and William F. Allman (eds.), *Newton at the Bat: The Science in Sports* (Scribner, 1984).
A collection of magazine articles on why Frisbees fly and how AstroTurf works.

For Grown-ups to Do With Children

Abruscato, Joe, and Jack Hassard, *The Whole Cosmos Catalog of Science Activities* (Goodyear, 1977).
Although this one is listed elsewhere as a book for older children, the excellent and varied activities, with a parent's help, are available to much younger children.

Arnold, Lois B., *Preparing Young Children for Science* (Schocken, 1980).
This guide for parents offers lots of suggestions on ways to nurture a child's natural search for order.

Brown, Sam Ed, *Bubbles, Rainbows and Worms* (Gryphon, 1981).
Activities for preschoolers that integrate science into everyday explorations of the world.

Goldberg, Lazar, *Children and Science* (Scribner, 1970).
This now out-of-print book, still available in many libraries, is a personal favorite—an intelligent, gentle, and inspired look at ways child and subject meet.

Hirsch, Elisabeth S., *The Block Book* (National Association for the Education of Young Children, 1984).
A firm believer in the value of blocks for children explains their potential import in perceptual development.

Holt, Bess-Gene, *Science with Young Children* (National Association for the Education of Young Children, 1977).
A guidebook for parents and teachers interested in providing thought-provoking activities for the very young.

Levenson, Elaine, *Teaching Children About Science: Ideas and Activities Every Parent and Science Teacher Can Use* (Prentice-Hall, 1985).
Objectives, background information, activities, procedures, and further resources useful for teaching scientific concepts to the very young. Typical units focus on the five senses, simple machines, weather, fossils, and ecology—there's lots here.

McIntyre, Margaret, *Early Childhood and Science* (National Science Teachers Association, 1984).

This collection of reprints from *Science and Children* presents both theoretical and practical information for parents as well as teachers.

Sprung, Barbara, *et al., What Will Happen If . . . Young Children & the Scientific Method* (Educational Equity Concepts, Inc., 440 Park Avenue South, New York, 1985).

Young children are introduced to the physical sciences by a group whose interest in gender equity is apparent.

Stronglin, Herb, *Science on a Shoestring* (Addison-Wesley, 1976).

A must for teachers, youth group leaders, and parents looking for ways to make science accessible. Here are low-cost, easy-to-find items and good explanations.

UNESCO, *New UNESCO Sourcebook on Science Teaching* (distributed in the U.S. by UNIPUB).

This enormous collection of activities in the physical, biological, earth, and space sciences is excellent. It comes in and out of print often, so check again if you can't find it now.

11 Wild Wisdom: Biology Out-of-Doors

Biology is the study of living systems. Each plant or animal has a natural life cycle; it uses resources and becomes a resource for other living things. Biologists are people interested in understanding that order and those interdependencies. They observe, classify, describe, and analyze plant and animal behavior in both natural and designed environments.

Although much of what biologists do is accessible to children, two powerful roadblocks keep young people from developing their skills and interests as amateurs. First, for children with no strategies to order what they see, all nature is an immemorable mass of color and activity—a forest is brown down low and green above, an anthill is sand crawling with thin black legs, and all deer are Bambi.

The other problem is attitudinal and equally serious. If potential observers see themselves as the only important or interesting species, then trees become what you carve your initials in, ants are what you step on at picnics, and deer are models for plastic statues to decorate lawns.

There are a number of things you can do to dismantle these roadblocks, many of which are described in the first half of this book. There are also items you can purchase or borrow to make observing more rewarding and satisfying.

HANDS-ON

Simply listing biology toys and equipment currently on the market makes it appear that "observing" is synonymous with "collecting," and collecting is fast translated as "hoarding." Children of all ages do well to understand the pleasures of unobtrusive observation. As a quiet, patient observer, equipped with a few simple tools, one is able to lift a curtain and peek into a startlingly magnificent world—a world not accessible to boisterous treasure seekers.

The tools of the observer vary from location to location. Snorkel gear allows for human entry into the undersea kingdoms; a large cardboard box set up as a blind offers the woodland naturalist close visual access to a nearby pond or stream.

Viewing Devices

Binoculars

Binoculars, handy in many situations, are a welcome gift. If you're in the market for a new pair, recognize the coding system used by manufacturers. The first number (for example, 7X) tells you how many times the object is magnified; the second number (for example, 35) refers to the size of the lens. The larger the magnification, the steadier one needs to be when holding the instrument; for 50X a tripod is recommended. Obviously the first number is important because it suggests how big a faraway object can be made to appear. At least as important as magnification, however, is lens size; for tracking animals on the run or birds which inconveniently hop from limb to limb, the wide lens is a considerable advantage. A 7 × 35 wide-angle lens is generally recommended.

Cost in binoculars is determined not only by the size of the lens but by the quality of the glass and the construction of the instrument as well. The better the lens, the less distortion in the image. What you see is what you get, so try them out before you buy. To save on cost, binocular manufacturers may glue rather than rivet lenses in place. The decision on whether the riveting is worth the cost depends on the intended use as well as personal finances.

Note: When inverted, binoculars can also serve as a makeshift field microscope.

Second Note: Quiet and patience are harder to come by than a pair of good binoculars.

Terrestrial Telescopes

Terrestrial telescopes use one eyepiece, whereas binoculars use two; they are less portable and often more difficult to use than binoculars. This may mean improved optics for less money. Terrestrial telescopes should differ from astronomical scopes in design. While no one cares if a star appears "upside-down,"an inverted moose is almost impossible to follow. Terrestrial scopes work well for people with a permanent nature blind, or for those who observe mostly from a stationary deck; for other purposes binoculars are generally recommended.

Feeding and Housing Equipment

By providing the wildlife in your area with a source of easy and desirable food, you can regularly bring animals into view. This works with cockroaches and bears as well as deer and birds: I recommend that you feed only those animals that are not dangerous to humans, and to whom you have an ongoing commitment.

Did you know, for instance, that birds go back to the same feeders year after year? If you want to attract lots, choose a food they like (sunflower seeds and suet are good bets), and get your supply out before your neighbor's appears. A loyal and fascinating clientele should soon arrive.

Bird feeders that hang directly on the window make close-up viewing a particular pleasure. Help children notice what a bird does to keep warm on cold days, how it turns its head to look for signs of danger, when it chirps and when it doesn't.

By providing food and shelter for life-forms, you can also help children understand natural cycles. For instance, once a bird has established a nest, children can occasionally peak in at the eggs and watch as the young offspring hungrily mature. Jed Burtt, an ornithologist, suggests that children put odd bits of string and yarn into a slotted strip of cardboard. In spring they can watch the birds collect housing materials, and in the fall, after the birds have abandoned their nests, they can search for spots where their yarn was used. Do certain birds prefer specific colors?

Although animal houses provide a site for action-viewing, there is a satisfaction in giving something as stable as a plant a home. Again, help children understand that plants respond to varying conditions.

Collecting Equipment, or Owning Is Not Knowing

Collecting, at its best, is a thoughtful activity. Children respond well to the suggestion that bugs can be caught, examined, and set free again. They are interested in the fact that a mayonnaise jar is not a suitable home for a beetle in temporary captivity, and that butterflies are injured by even the most gentle thrust of a butterfly net. Most find it fascinating that an empty seashell was once the home of something living and that, if left in place, may again become a suitable spot for a hermit crab or insect to occupy.

Collectors should learn to consider the uniqueness of the species they are about to grab. If a hundred people came tomorrow to take a similar leaf, would there still be plenty left? If the answer is yes, take one for your collection. If it is no, map the location of the plant; keep a diary of the species' availability, and check the same spot next year. Are there more or less of the rare breed? A notebook or an artist's sketch pad can serve the collector better than a shoe box full of items.

In addition, be sure that children understand the purpose of a biological collection, which is to learn more about a particular environment. To this end, all samples should be identified by date, location, and species. Collectors, like all naturalists, must have access to field guides.

Gifts for Collectors

Leaf and flower presses are wooden frames layered with blotter paper and screwed down tight with wing nuts. Leaves or flowers left in this contraption for two weeks retain their color and shape, and emerge ready for the scrapbook.

Fig. 8. An attractive, well-made and effective flower press is available from Sunstone (P.O. Box 788CS, Cooperstown, NY 13326) for approximately $10—a "best buy" on flower presses.

Fig. 8

THE GETTER is a clear plastic tool for catching insects, painlessly ejecting a spider from inside the house, or removing newborn guppies from a tank. Available from *Toys to Grow On* and pet stores for approximately $4.

THE BUG HOUSE, a $6 wire mesh device which gives captured insects sufficient air, and the young entomologist an unbreakable container with no sharp edges.

Fig. 9. **THE GETTER, BUG HOUSE,** a fine two-way viewer, and a book about insects—the perfect beginner's kit—are sold as a package from *Toys to Grow On* for $16.

Fig. 10. **BUG BOXES**—actually a misnomer for these plastic magnifying containers that work as well for stones, stamps, or pennies as for bugs—sell for less than $1; these are great stocking stuffers. Available at museum stores and hobby shops.

Games

For All Ages

YOTTA is a perfect game for children of all ages to practice animal identification. Excellent color photos are set out on a square grid, and by a toss of the dice, players are directed to name a beast. The gameboard is well made, and

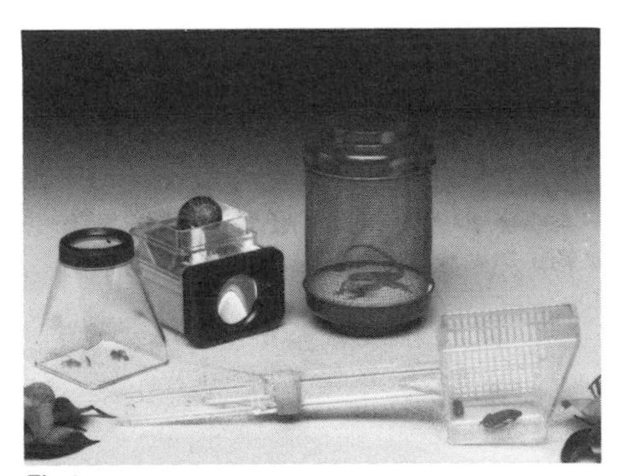

Fig. 9

Fig. 10

additional sets of cards can be purchased: Birds (88 species), Waterfowl (54 species), and Mammals (63 species). Cards and gameboard retail for approximately $14 at better toy shops, museum stores, and through Carolina Biological. Additional cards are sold separately for $7. Highly recommended.

Animal rummy cards, beautifully photographed decks from Safari Ltd. which sell for less than $2.50 each, can be used with young children as "flash cards" and with older kids as the rummy game for which they were intended. Available sets include Endangered Species, Snakes, Mysteries of the Deep, Zoo Animals, and Hidden Kingdom Insects.

For Young Children

There are a number animal replicas, lotto games, puppets, card games, puzzles, stickers, rub-ons, and identification books available in toy stores, drugstores and toy supermarkets. Zoologists are generally appalled: "The ears are way out of proportion" or "Those fins are in the wrong place."

I am not as put off by this motley collection of beasts. Young children do not go to a jungle puzzle, for instance, to comment on the flora and fauna in the background; in most instances the picture is much less important than fitting the pieces together. Furthermore, for someone unacquainted with the animal forms, there is a general correspondence between the actual beast and the one depicted. On the other hand, I do not understand why designers of such items cannot be more precise in their renderings. It is just as much fun to rub off a decal which is accurately colored and properly identified.

If you are interested in improving your child's powers of observation, ask him/her to compare the likeness to the actual beast (sure, take the sticker to the zoo), or at least to an accurate rendering in an encyclopedia or biology book. Most of these "generic animals" do not inspire science questions or serve as the basis of reasoned speculation. In short, if you have a choice, go for the accurate rendering.

Fig. 11

Fig. 11. **Life cycles** puzzles, designed for ages three to nine, are an example of the scientifically accurate, moderately priced items recommended here for young children. For under $3 you can purchase The Butterfly, The Frog, The Apple, The Robin, The Chicken, or The Honeybee.

For Grade-Schoolers

ENDANGERED SPECIES ($12) is a board game played much like **MONOPOLY**, but instead of buying hotels for Boardwalk, a player buys preserves for endangered animals. As one saves a monkey-eating eagle from hunters and zookeepers, or the sea otter from chemical pollutants, the sometimes hopeless plights of these species becomes clear. The graphics are well done, and descriptions accurate. Unfortunately, knowing something about the animals is no advantage in this game, just as a knowledge of real estate is not essential to a win in MONOPOLY. Recommended for age nine and older.

PREDATOR is a relatively slow game based on a monochromatic deck of animal, plant, and "death and decay" cards. An animal takes what it eats and is taken by what it eats. Plants, as basic producers of food, score higher energy points than animals. Recommended for children age eight and older who are out to learn, this one is available for $5.75 direct from Ampersand Press, 2603 Grove St., Oakland, CA 94612.

POLLINATION, based on a deck of seventy-two colorful cards, teaches children to match floral species with the right pollinators; moves are determined by the roll of the dice. Also from Ampersand Press (see **PREDATOR**), for $6.50.

For Middle-Schoolers

KRILL is a food-chain card game set in the Antarctic Ocean. Since a decision about harvesting krill for human consumption will be made in the next few years, the relationship of this species to other marine life is particularly important to understand. The game itself is a bit cumbersome, but the fact that it can be played in large or small groups may offset this criticism. From Ampersand Press (see **PREDATOR**); $6.25.

IN PRINT

Field Guides

Books designed to help naturalists identify what they see are called field guides. Guides to trees, birds, mushrooms, ferns, seashells, wildflowers, weeds—almost any natural form you can name—are currently available. Begin by purchasing *one* guide which focuses attention on your immediate surroundings; overwhelming a kid with too much information is counterproductive. Wildflowers, which can be identified by color, are easier to learn than ferns; for the same reason, birds are easier to learn than insects. Consider also the aesthetic tastes and daily habits of the potential user—for a color-blind kid who can't wake up in the morning, birds are a poor choice.

Purchase a guide that the user can handle. The Peterson Field Guide series (Houghton Mifflin) and the Audubon Society Field Guide series (Knopf) though designed for adults can be used by young people. The Golden series also gets good reviews from naturalists. Keith Rushforth's *Pocket Guide to Trees* (Simon & Schuster, 1981) is considered a model work of this kind, accurate and easy to use.

An altogether new concept in field guides is exemplified by Bill Perry's *A Sierra Club Naturalist's Guide: The Middle Atlantic Coast* (Sierra Club, 1985). This volume contains identification information on both the flower and fauna of a region so that one needn't pack five books to help recognize wildflowers, trees, birds, ferns, and insects. Other volumes in this series focus on the North Atlantic Coast, Southern New England, the Piedmont, the North Woods, the Sierra Nevada, and the deserts of the Southwest.

Begin working with your child to memorize some of the most common species in your area. Recognizing Queen Anne's lace, for instance, not only allows for the momentary thrill of identification, but might also enable a youngster to identify more than half the flowers in the field behind the house. Once a specimen is recognized in full bloom, the child can find examples of the flower before and after its peak. It is just as important for them to notice that Queen Anne's lace grows in open fields, and to look at the quality of the soil surrounding the plant, and to understand that this flower is host to a variety of insects, as it is to learn name after name.

Coloring Books

At their best, coloring books invite children to look more carefully at the world, noting the shapes of insects, the plummage of birds, the variations evident in a single species of wildflower. The following are recommended for *older* children. Younger ones will be frustrated by them and appropriately distressed by your attempts to insure that they "get it right."

Fig. 12

A series of coloring books based on the respected Peterson Field Guides are available from Dover Press for $4 each. These fine line drawings teach children (and adults) about wildflowers or birds, as they produce the best "paint-by-number" works to date.

The Entomological Society of America also produces a highly reliable insect coloring book for children, again using the color-by-number technique. This one can be purchased for $3 direct from E.S.A., Box 4104, Hyattsville, MD 20781. Maryland residents add 5% sales tax. Multiple-copy discount available.

Also in print are coloring books based on *Gray's Anatomy* ($5, Running Press, 1980), Audubon's *Birds of America* ($2, Dover, 1974), and Thomas M. Niesen's renditions of marine life ($9, Barnes & Noble, 1982). For sophisticated dinosaur fans, try Antony Rao's *Dinosaurs,* which features the beasts in chronological order (Dover, 1980).

If you object to the mechanistic aspect of "color-by-number," take out a field guide and let your children choose their own colors based on the variations described in the source book. Field guides also provide a great answer to the question commonly asked by young artists: "What does a flower [or a bird or a tree] look like?"

Wearables

Clothes, at their best, are a celebration of what you value and enjoy. Here's what the fashionable biology fan is wearing these days.

Fig. 12. ENDANGERED SPECIES T-shirt ($13) and sweatshirt ($17) kits from Nancy's Kreative Designs (2686 Caminito Prado, La Jolla, CA 92037) come with a silk-screened Emperor Penguin, Bald Eagle, or Siberian Tiger and paints. Children are proud to paint their own clothes, and they learn something about the species from the documentation provided. Moreover, Nancy Dinenberg, their creator, donates a portion of the proceeds to the National Wildlife Foundation. Satisfying, educational, and fun.

Fig. 13

What may well be the world's largest collection of scientific-technical-humorous T-shirts can be previewed in a catalog from Software Unlimited International. Send 40 cents or two postage stamps to Software Unlimited, Route 2, Box 444B; Twisp, WA 98856.

For more extraordinary T-shirts see the catalog from the Nature Company.

Fig. 13. Anatomy T-shirts from Nasco come in kids' sizes and sell for $9 to $11.

An electron microscopist gave her line of lab coats, T-shirts, dresses, posters, and postcards featuring DNA and RNA magnified 35,000 times the name **DESIGNERGENES.** Sure to inspire questions like "What is *that?*", your kid can be both fashionable and in the know ($6 to $20).

Posters

Wildlife posters are yet another way to celebrate life on this planet. Many museums and zoos—the Steinhart Aquarium, San Diego Zoo, California Academy of Sciences—have commissioned prints worthy of anyone's living room. These can be purchased directly from these institutions or from The Nature Company, which carries an excellent selection of prints.

The Sierra Club also sells a spectacular series of wildlife prints for about $18 each.

Accurate and beautiful murals from the Peabody Museum of Natural History at Yale are sold through Carolina Biological; the "Age of the Reptiles" and the "Age of the Mammals" come in a colorful 9′3″ × 20″ panel and sell for $50 each.

The U.S. Government Printing Office produces marine life posters which are both beautiful and relatively inexpensive. These include:

Marine Fishes of the North Atlantic, S/N 003-020-00027-4, $5.50
Fishes of the Great Lakes, S/N 003-020-00069-0, $5.50
Marine Fishes of the California Current, S/N 003-020-00055-0, $5.50
Marine Fishes of the North Pacific, S/N 003-020-00051-7, $5.50
Sea Turtles of the World (36″ × 24″), S/N 003-020-00152-1, $5

Nature and Ecology Organizations

There are a number of national organizations dedicated to the study and preservation of the environment. They are good sources of additional information and are candidates for your support. Among the largest of these are:

The Cousteau Society, Inc., 8430 Santa Monica Blvd., Los Angeles, CA 90096. Concerned with the seas, as other organizations are with land, the Cousteau Society publishes an excellent little magazine for their child members, *Dolphin Log.*

The Fund for Animals, 200 West 57th St., New York, NY 10019. Their motto, "We speak for those who can't," refers to animals, both domesticated and wild, that are endangered.

The Humane Society of the U.S., 2100 "L" St., N.W., Washington, DC 20037. Organized to "promote the humane treatment of animals and to instill kindness in mankind." Publishes a juvenile periodical, *Kind,* as well as other relevant information.

The National Audubon Society, 950 Third Ave., New York, NY 10022. Interested in ecology, energy, and the conservation and restoration of natural resources, with an emphasis on wildlife. Numerous education programs and materials, as well as summer camping opportunities.

The National Wildlife Federation, 1412 16th St., N.W., Washington, DC 20036. Interested in the management and appreciation for natural resources. Provides grants for conservation projects, sponsors National Wildlife Week, and publishes children's magazines *Your Big Backyard* and *Ranger Rick.* Also sponsors Ranger Rick Nature Clubs and provides inexpensive or free educational materials.

The Sierra Club, 530 Bush St., San Francisco, CA 94108. Concerned with nature and its interrelationship with humans, this is one of the largest and most active lobbyist and educational groups in the conservation movement.

Sports Fish: reproductions of ten wildlife paintings by Bob Hines, 17″ × 14″ each, sold only as a set, S/N 024-010-00277-8, $6

Magazines

For Young Children

Chickadee, 10/year, $14 in U.S., $12 in Canada (Young Naturalist Foundation, 59 Front St., Toronto, Ontario M5E 1B3, Canada). This Canadian magazine for age eight and younger is designed to interest children in the world around them. The format is excellent, the articles well written, and games and puzzles appealing.

Tips for Observing

- Recognize that animals move fast; for the easily frustrated beginner, botanical forms are easier to identify and study.
- If a child wants to study animals, begin with tracks, or by helping him/her notice features which are easily recognized and classified. Describing size, color, location, and movement patterns takes practice.
- Find ways to limit the field of view. A former science teacher suggests bending a hanger into a diamond shape, placing it on the ground, and having children map all that appears in the marked territory. Using one's hands as binoculars also helps in this way.
- Don't forget window bird feeders, wildflower gardens, stocked ponds, and salt licks. They bring nature closer and give children time to watch.
- If possible, study the behavior of one plant, insect, or vertebrate; all sparrows, for instance, are not alike. Use nail polish to paint a colorful dot on a tiger beetle, release it, and keep notes on its activities.
- Become familiar enough with an area—a backyard, a park, a neighborhood—so that day-to-day and seasonal differences are obvious. Try to account for the changes observed.
- Go out in a creative mood. Create alternative explanations. "How did it do that?"

Your Big Backyard, 12/year, $8.50 (1412 16th St., N.W., Washington, DC 20036). This monthly publication of the National Wildlife Federation, designed for children ages three to five, features splendid color photos, simple descriptive text, games, puzzles, and a letter for parents.

For Older Children

Animal Kingdom, 6/year, $8 (New York Zoological Society, Zoological Park, 185 St. and Southern Blvd., Bronx, NY 10460). A publication for children in grades 7 and up, describing the habits and behavior of animals around the globe. Good black-and-white photos and illustrations add much to this well-written and -conceived periodical.

Audubon, 6/year, $13 (National Audubon Society, 950 Third Ave., New York, NY 10022). Well-written and lavishly illustrated articles on the preservation and habitat of native species. For readers in junior high and above.

The Curious Naturalist, 4/year, $4 (Massachusetts Audubon Society, Lincoln, MA 01773). A nature publication for families from a group with an excellent track record. Articles and activities invite a better understanding and appreciation of stars, flowers, birds, water creatures, and more.

Dolphin Log, 4/year, $10 (This includes membership in the Cousteau Society, 8430 Santa Monica Blvd., Los Angeles, CA 90069). This educational publication for children ages seven to fifteen includes articles and games about marine biology, ecology, the environment, and natural history. Well written and researched.

National Wildlife, 6/year, $12 (National Wildlife Federation, 1412 16th St., N.W., Washington, DC 20036). This magazine for children in grades 6 and above is designed to encourage the wise and proper use of those resources our lives depend upon—soil, air, water, minerals, plants, and wildlife. A colorful and intelligent publication.

Ranger Rick's Nature Magazine, 12/year, $12 (National Wildlife Federation, 1412 16th St., N.W., Washington, DC 20036). Perhaps the best-known of the juvenile nature periodicals, this magazine for children ages six to twelve includes both indoor and outside activities to increase their appreciation and understanding of conservation. The approach varies; stories, plays, informational articles, stunning illustrations, and puzzles encourage children in a lively search for order and meaning in their world.

Informational Literature

The books in this list describe ideas and projects that can be appreciated by the young and not-so-young alike.

For All Ages

Arnosky, Jim, *Secrets of a Wildlife Watcher* (Lothrop, Lee & Shepard, 1983).

A four-star book on techniques for observing animals in the wild. Highly recommended.

Mabey, Richard, *Oak and Co.* (Greenwillow, 1983).

This book tells of the two-hundred years in which insects, birds, and animals interact with a stately oak. Beautifully illustrated.

Newton, James R., *A Forest Is Reborn* (Crowell, 1982).
After a destructive fire, life begins to stir again in the forest. Newton's easy-to-read text and Bonner's evocative illustrations explain the process.

Overbeck, Cynthia, *Sunflowers* (Lerner, 1981).
Startling color photographs and simple but detailed descriptions of the life cycle of a common plant.

Oxford Scientific Films, *Harvest Mouse* (Putnam, 1982).
Just one in a series of spectacular books of colorful, larger-than-life photos. Basic information is presented in a four-page introduction, and the captions, written for children of all ages, tell the rest.

Parnall, Peter, *The Daywatchers* (Macmillan, 1984).
A sensual and artistic look at birds of prey from an experienced naturalist.

Rights, Mollie, *Beastly Neighbors: All about Wild Things in the City* (Little, Brown, 1981).
Generally recommended for older children; experience suggests that experiments suggested here make excellent family projects.

Stewart, John, *Elephant School* (Pantheon, 1982).
Elephant nature is evident as a boy mahout in Lampang, Thailand, trains this huge mammal. A superb photo essay.

Tveten, John L., *Exploring the Bayous* (McKay, 1979).
If this one doesn't inspire you to pack for a trip to Louisiana or Texas, nothing will.

Yount, Lisa, *Too Hot, Too Cold, Just Right: How Animals Control Their Temperatures* (Walker, 1981).
Here's the well-explained answer to a fascinating question.

For Young Children

Arnold, Caroline, *Animals That Migrate* (Carolrhoda, 1982).
Amazing facts, clear explanations, and sensible presentation make this one a good choice for young readers.

Isenbart, Hans-Heinrich, *Baby Animals on the Farm* (Putnam, 1984).
Superlative color photos highlight an informative text.

Miller, Edna, *Mousekin's Mystery* (Prentice-Hall, 1983).
One in a series of amazing Mousekin picture books, in which the world is described from the small animal's perspective. Here Mousekin encounters foxfire and natural bioluminescence.

Ryder, Joanne, *The Snail's Spell* (Warne, 1982).
"Imagine that you are soft and have no bones inside you. . . ." This artistic and imaginative trip inside a snail is hard to resist.

Selsam, Millicent, and Ronald Goor, *Backyard Insects* (Four Winds, 1981).
The intriguing ways insects camouflage themselves is revealed through a brief text and marvelous close-up photos.

Skofield, James, *All Wet! All Wet!* (Harper & Row, 1984).
A small boy unobtrusively observes woodland creatures on a rainy day. A lyrical and scientifically accurate picture book.

Wolff, Ashley, *A Year of Birds* (Dodd, Mead, 1984).
Each month different birds visit Elsie's feeder and yard. Block prints imaginatively show seasonal changes.

Yabuuchi, Masayuki, *Whose Baby?* (Philomel, 1985) and *Whose Footprints?* (Philomel, 1985).
Two superb examples of nonfiction for young children, these books invite readers to classify and describe. Colorful photographs, rich in detail, add immeasurably to their task.

Zweifel, Frances, *Animal Baby-Sitters* (Morrow, 1981).
Range cows, elephants, acorn woodpeckers, and macaques are four animals that live in herds and use mother's helpers. And you thought you were the only ones!

For Grade-Schoolers

Arnold, Caroline, *Saving the Peregrine Falcon* (Carolrhoda, 1985).
An endangered species may be saved from extinction by scientists who are raising the bird's fragile eggs—eggs which would not survive in the wilderness. Color photos and moving text that highlight their efforts.

George, Jean Craighead, *One Day in the Desert* (Crowell, 1983).
The desert ecosystem explained in George's memorable style.

Goor, Ron and Nancy, *All Kinds of Feet* (Crowell, 1984).
A variety of feet are explored in this excellent introduction to comparative anatomy.

Lauber, Patricia, *Seeds: Pop—Stick—Glide* (Crown, 1980).
An award-winning title which focuses on the most common of flowers, the dandelion.

MacClintock, Dorcas, *A Raccoon's First Year* (Scribner, 1982).
A weekly chronicle, illustrated with black-and-white photographs, of the raccoon's development, anatomy, and behavior.

Seuling, Barbara, *Elephants Can't Jump and Other Freaky Facts about Animals* (Lodestar, 1984).
An entertaining look at the animal kingdom.

For Middle-Schoolers

Arnosky, Jim, *Drawing from Nature* (Lothrop, 1982).
For artist and non-artist alike, this one helps you notice nature in surprising detail.

Attenborough, David, *Discovering Life on Earth* (Little, Brown, 1981).
Good for reference or browsing, this profusely illustrated chronological summary of life is readable, informative, and entertaining.

De Beer, Gavin (ed.), *Charles Darwin and T. H. Huxley* (Oxford, 1984).
This inexpensive paperback contains two fascinating autobiographies designed to give older children a sense of Victorian science and culture.

Hughey, Pat, *Scavengers and Decomposers: The Cleanup Crew* (Atheneum, 1984).
The process of biodegradation is clearly explained in this story of natural interdependence.

Johnson, Sylvia, *Wasps* (Lerner, 1984).
Remarkable close-up photos make this discussion of metamorphosis, behavior, and role differentiation exceptional.

Kohl, Judith and Herbert, *Pack, Band and Colony: The World of Social Animals* (Farrar, Straus & Giroux, 1983).
A discussion of wolves, lemurs, and termites which clearly explains the social nature of animals.

National Geographic Society, *Secrets of Animal Survival* (National Geographic Society, 1983).
The color photos that make *National Geographic* famous are used here to describe animal adaptations.

Peacock, Howard, *The Big Thicket of Texas: America's Ecological Wonder* (Little, Brown, 1984).
Eight major ecosystems are found on 3.5 million unique acres of eastern Texas. Peacock decribes the flora and fauna, the history of the area, and the battle of conservationists trying to keep it intact.

Riedman, Sarah R., *Biological Clocks* (Crowell, 1982).
Plants, animals, and people are all affected by nature's rhythms. Riedman tells how.

Scott, Jack Denton, *The Fur Seals of Pribilof* (Putnam, 1983).
These mammals travel 6000 miles each year to and from Pribilof, where they give birth and mate. Scott describes the event in a wondrous photo essay.

Selsam, Millicent, *Tree Flowers* (Morrow, 1984).
Detailed color drawings of a flowering tree face a page of relevant botanical and historical information. A beautifully designed, accurate book.

Spruch, Grace Marmor, *Such Agreeable Friends: Life with a Remarkable Group of Urban Squirrels* (Morrow, 1984).
A perfect tribute of the individuality of animals, and to the powers of observation cultivated by a careful non-scientist.

Taylor, Ron, *The Story of Evolution* (Warwick, 1981).
Heredity, survival, adaptation, and more are discussed in this important text.

For Parents

Allen, Sarah (ed)., *Explorer's Notebook* Series (Little, Brown, 1980).
These three separate guides to the insects, birds, and flora of the eastern U.S. help children recognize species and take scientifically interesting notes on what they see.

Chinery, Michael, *Enjoying Nature with Your Family* (Crown, 1977).
Lots of projects and experiments, as well as a reference section on places to visit and organizations concerned with natural history.

Cornell, Joseph Bharat, *Sharing Nature with Children* (Ananda, Nevada City, CA, 1979).
Forty-two nature awareness games for kids and adults.

Horwitz, Eleanor, *Ways of Wildlife* (Citation, 1977).
Games, field trip ideas, stories, and other activities designed to teach children about the relatedness of all things. An informed and informative text.

Sisson, Edith A., *Nature with Children of All Ages* (Prentice-Hall, 1982).
If I were to buy only one book on nature activities for children, this would be it. Intelligent, lively suggestions from an experienced children's guide.

12 Magnifiers and Microscopes: The Better to See You With, My Dear!

As any kid who's spent an afternoon looking through a soda pop bottle can tell you, a curved piece of glass distorts an image. Magnifiers and microscopes work on this same principle: a curved piece of glass or plastic is set in a rim, and items viewed through it appear bigger. They help users differentiate between a fossil and a scratch, enable a biologist to grab the right piece of tissue with tweezers, and allow a young child to understand that cloth is, in fact, woven thread. Using a compound microscope one can see objects and structures simply not visible to the naked eye.

HANDS-ON

Magnifiers

Ideally, in the drawer where the family scissors, tape, and stapler are kept, one finds a magnifier. The size, quality, and shape of the instrument should reflect the needs and purposes of the household. A lighted glass with a stand is appropriate for stamp and coin collectors, a lightweight hand lens is suitable for fossil hunters, and an oblong instrument works well for those needing help with small print. I judge the success of an instrument by the frequency with which it is used.

Treat the magnifier as an important tool. Help a child study a flower's stamen, or dust from a table and compare it to dust from the windowsill. Look at the shape of quartz crystals, and the patterns of finger and toe prints. Take the magnifier out to repair a piece of jewelry or an electronic part.

There are few gifts more useful than a portable, easily focused magnifier which enlarges objects 4 to 6X (times). Such an instrument helps focus attention on a particular area and teaches early the importance of a tool. Before purchasing one, consider the following:

- Magnifiers vary in size. For young children, purchase an instrument with a lens diameter of at least 3″, a lens large enough for them to see a crawling insect before it crawls away.
- Magnifiers vary in weight. You want an instrument that won't tire youngsters out.
- Focusing a magnifier (that is, moving it up and down) is often difficult for young kids. Consider purchasing a magnifier on a tripod, or a lens like the TWO-WAY MICROSCOPE (see illustration on page 155) in a free-standing plastic collar.
- The quality of lenses varies. A ground-glass lens is heavier, but causes less distortion; a plastic lens is less expensive, but more easily scratched.
- Safety and durability are of concern. Ask yourself:

  (a) Can the lens fall out?

  (b) Can the handle be unscrewed, leaving a small child with a sharp point and no handle?

  (c) Are there sharp edges anywhere on the device?

  (d) Will the lens shatter or crack?

- Your child's interests and track record are factors. If the kid loses things, find an instrument with a hole in the handle, and tie it to the child's belt. If tromping in weeds is his/her forte, don't purchase a magnifier with a large tripod.

Fig. 14

Fig. 15

Fig. 14. Magnifiers come in a variety of shapes and sizes. Consider potential use before purchasing one.

Fig. 15. Giant magnifiers usually contain a 4″ or 5″ prefocused lens and sell for $20 to $30. Recommended for preschoolers, these devices let more than one child look at the same time.

Fig. 16

Fig. 16. A four-star, inexpensive ($6) magnifier called the TWO-WAY MICROSCOPE, distributed by Suitcase Science (and Battat), is available in toy stores and museum shops. The prefocused lightweight lens sits on a plastic stage; a second eyepiece and angled mirror allow for underneath viewing. Recommended by children and educators alike.

Microscopes

There are times, however, when one wants to see more detail than a 6X lens provides. A magnifier designed to enlarge objects 30X, for instance, would be hopelessly difficult to produce; the large and unwieldly piece of glass would distort objects beyond recognition. Optics people have developed a neat solution to this problem, however. By placing one lens on top of another, the user is actually able to multiply the magnification, one lens times another. This is the design upon which the compound microscope is based. The microscope's tube holds the two lenses in place and is used to direct the light one needs in order to view any object.

Naive logic suggests that the more powerful the microscope, the more the user sees. Thus, the average shopper looks for the greatest degree of power (with some consideration given to the quality of the lens) for the least money. In the same way that hot-rodders brag about power under the hood, academically oriented kids compare microscopes: "I've got 1200X."

But microscopists understand the situation differently. Those in the know insist that degree of magnification is only one of several variables purchasers need to consider; resolution (the clarity of the image) and the accessibility of light are equally important. The problem from the consumer's point of view is that there is no easy way to say on the package "This device produces images that are twice as clear as that device."

Think of it this way: Imagine a radio that is so powerful it can pick up stations anywhere, and each station, so as not to interfere with any other, has established a precise code for tuning in. Tuscaloosa, Alabama, for instance, is 87.26843 on your radio dial. Now pretend to tune in Tuscaloosa on your conventional radio. Because there are so many stations in between 87.1 and 87.3, you get nothing but loud static. Your instrument is simply ineffective.

A similar phenomenon occurs with an overpowered microscope. Using a 1200X scope, for instance, even the slightest movement takes you past the point you wish to see. Sophisticated instruments compensate for this problem by including a fine-focus device, referred to as a fine-focus wheel, but again, the higher the power, the better the fine focus has to be. A typical college lab microscope, which sells for $300 to $400, includes a maximum of 450X. In sum, it is not particularly difficult to put two decent lenses in a tube—the hard part is designing an effective instrument.

But even if you could purchase an excellent instrument which magnified items 1200 times, I wouldn't recommend it. While at 30X you can see little vinegar eels swimming on a slide and count how many are active and alive, with a high-powered instrument none of that activity is visible—the field of vision is simply too narrow. It is as if by using boulders someone had written a message in huge letters that can only be seen from a helicopter: to the person standing a couple of inches away, one of the rocks used to form the letters is simply a stray boulder.

Moreover, slides for a high-powered scope are extremely difficult to make. While at 30X you can sprinkle salt on a piece of clear tape and see mountain-like crystals in amazing relief, with a 1200X instrument your tape would appear as nothing but a dark blob. Although one can purchase interesting prepared slides for a 1200X device, this is like watching someone else's travelog rather than enjoying your own snapshots.

And finally, since you are seeing only a tiny area with a high-powered microscope, the amount of light hitting that spot is also tiny. Thus, the image generally appears dark and featureless unless an extremely good light source is also available.

So . . . Rule Number 1: *Do not buy power.* I am generally suspicious of anything over 300X. The distressing part of this story is that toy manufacturers understand the naive desire for high magnification and cater to consumers' mistaken assumptions.

"How can you justify that?" I ask otherwise reputable sales representatives.

"Low-powered scopes will never sell" is their unequivocal reply.

Recommended Beginning Microscopes

A child beginning work with a microscope is best served by a device which magnifies objects 30 to 150 times their normal size. Four highly recommended items are currently on the market. The first is known as a "pocketscope," which most often looks like a 6″ × 2″ × 1″ plastic box with a peephole. A single lens magnifies items 25 to 35X, and a built-in light (sometimes with color-corrected filters) makes possible the viewing of opaque objects. Before purchasing a pocketscope, try it out; what you see is what you get. I recommend pocketscopes for the following reasons:

- They allow a user to examine anything—chicken bones, metal fittings, onion skins—and it all looks amazingly interesting.
- These devices are easy to focus.
- They are lightweight and portable.
- They help users decide what other kinds of things they might wish to view—whether the kid is more interested in counting swimming pond creatures or understanding more about the cell structure of a plant. Specific interests can determine which, if any, microscope they need next.

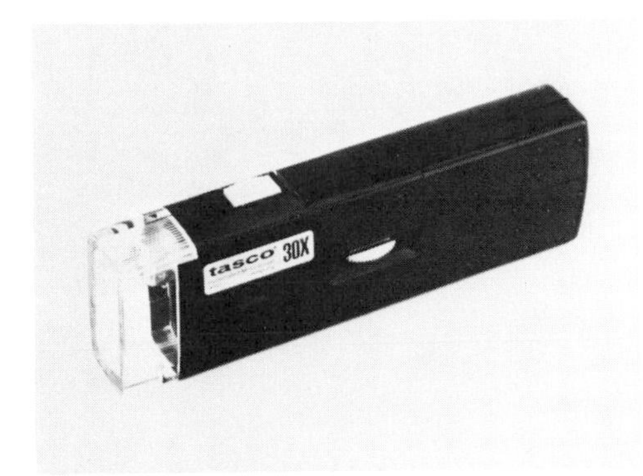

Fig. 17

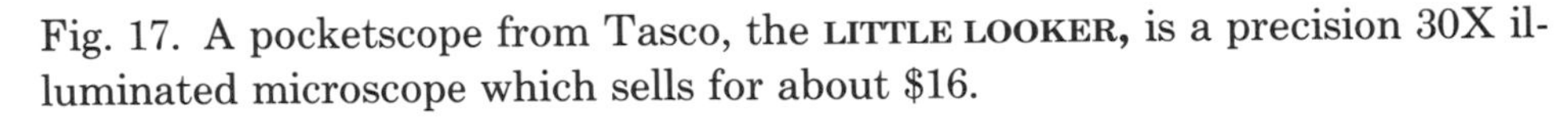

Fig. 17. A pocketscope from Tasco, the LITTLE LOOKER, is a precision 30X illuminated microscope which sells for about $16.

Also highly recommended is the SUPERMICROSCOPE from Learning Things, an instrument specifically designed to serve the needs of beginners. It is made of highly durable plastic, includes a helical focusing device, and is versatile and inexpensive (approximately $20). This 30X scope can be increased to 120X by snapping in an ocular lens/eyepiece. When children leave for field or stream, they simply snap off the mirror and the device becomes a nifty field microscope which is held up to an available source of light.

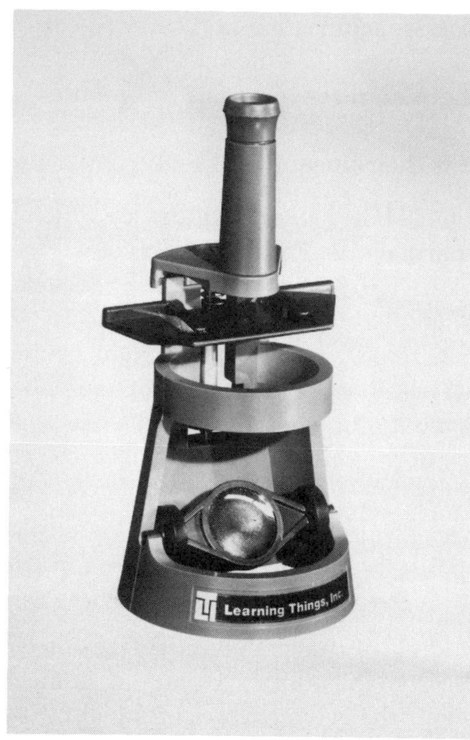

Fig. 18

(Learning Things warns you never to point the viewer directly at the sun, and they're *not* kidding!) With a mask attachment, the SUPERMICROSCOPE slips onto the lens of a slide projector and works beautifully as a projection scope. Although this three-in-one device looks less "scientific" than others in the under-$70 price range, it works better.

Fig. 18. The **SUPERMICROSCOPE** can be purchased alone or in a boxed kit which includes slides, bug box, pipet, and more. Available through science catalogs, in museum shops, or direct from Learning Things.

The third easy-to-use microscope, called the G.S.S. BLISTER VIEWER, is an all-metal 50X microscope with built-in light; a 100X objective can be purchased separately. This substantial instrument, manufactured in Minnesota, is well designed and wonderfully inexpensive (less than $35 when purchased direct). Blister slides made especially for this viewer are also easy to use and bargain priced. Write: General Science Service, P.O. Box 2022, Elbow Lake, MN 56531, or see the G.S.S. BLISTER VIEWER at places like the Exploratorium gift shop.

Fig. 19. The **G.S.S. BLISTER VIEWER** is an all-metal microscope with built-in light *above* the stage. A well-designed, substantial instrument.

Fig. 20. The **EXPLORERSCOPE,** a cardboard microscope which is focused by

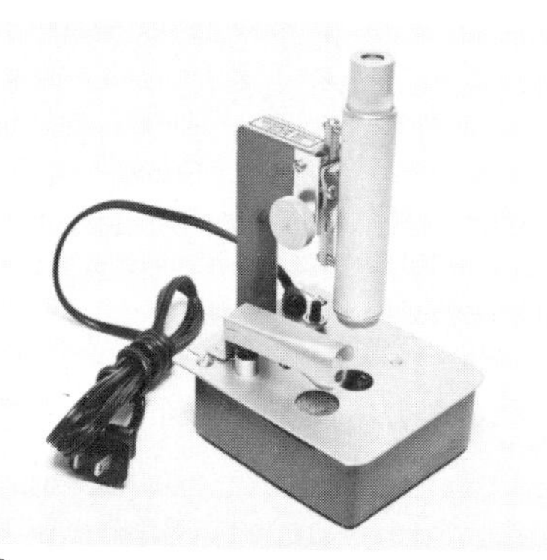

Fig. 19

Fig. 20

pressing the cardboard case, does the job of magnifying objects 30X extremely well. Recommended for children ages ten and older because focusing requires some dexterity, this $2.50 item is one of the best "stocking stuffers" of all times. Available in museum and science shops or direct from Learning Things.

Fig. 21. Magnifiers come in a variety of shapes and sizes. In some, lenses can be overlapped to produce an even larger image. These particular devices from Learning Things retail for under $2.

More Advanced Microscopes

Children who have had experience with less sophisticated instruments will eventually define their interests. For instance, if your young person wants to study cell structure and learn to make slides, and can plan several projects in advance, purchase a compound microscope like those used in high schools or colleges.

Quality instruments can be purchased new through science catalog houses (Nasco, Sargent Welch, Carolina Biological, etc.) for approximately $250. If these are too costly, investigate the availability of used microscopes in your area. An ad in a local paper may turn up doctors or dentists now ready to part with their old microscopes. Also check with local high schools or colleges; frequently they replace perfectly adequate equipment with "state-of-the-art" technology. Finally, in most city telephone directories there is a listing for "microscopes—used." Companies that specialize in buying and refurbishing instruments advertise here. Good-quality microscopes at reasonable prices can often be located in these ways.

If, on the other hand, your amateur biologist is more interested in viewing opaque objects—dissections, studying fruit fly mutations, or paramecia—consider buying a binocular scope, also called a stereo microscope. In these instruments the light comes from above the stage, and there is a greater distance between the stage and the objective lens. The most noticeable feature of the binocular system is the twin eyepiece; it makes extended viewing easier and produces a stereoscopic effect with real depth of field. Stereomicroscopes also work exceptionally well for beginners.

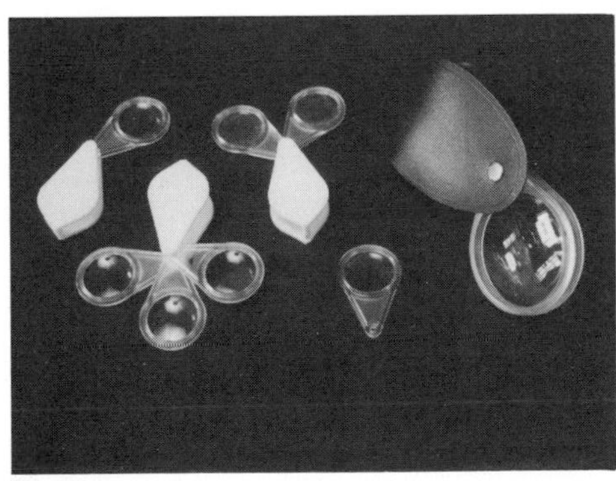

Fig. 21

Before Buying a Microscope: A Consumer's Checklist

√ Decide what you want to do with your microscope. Are you better off with a pocketscope, a versatile, durable item like the SUPERMICROSCOPE, a dedicated compound microscope, or a dissecting scope? One child may think portability extraordinarily important; another may be totally committed to a long-range and intensive study of fruit flies.

√ Determine the kind of power and quality of lens you need to see what you are interested in. Are your expectations realistic, given the money you have to spend on a device? If you cannot afford the microscope of your child's dreams, point out that specialized science equipment is not always something that you own. See if there are microscopes to borrow at school or in the library. Postpone your experiments, but *don't* buy an item which you know is inadequate.

√ Look at the overall design. Does the microscope feel solid and steady? Are there clips to hold the slide in place? Do any of the parts look as if they will break? For some strange reason many poor microscopes have a tilted or tiltable stage, an unnecessary feature which often makes viewing more difficult.

√ Understand that the amount of light coming through the hole is an important variable. Is the mirror easy to aim? A concave mirror is better than a flat one. Does the microscope come with a built-in light source? (Some people say that batteries may leak and corrode and ruin contacts, and that you are better off using a small lamp with a 50-watt bulb.) Also look under the stage—most good scopes have a click-stop diaphragm which turns to let in more or less light. Make sure that the diaphragm and filters aren't simply ornamental. Good filters may help the viewer see structures and behavior not clearly seen in white light.

√ Look at the focusing device. Is it helical or rack-and-pinion (the kind that looks like a gear climbing a notched pole)? Helical works better on nonprofessional scopes. Is there a fine-focusing device? Does the focus move too easily? You want your object to stay in focus once you get it there. Some scopes have a tightness adjuster built in. Also, are the lenses parfocal, that is, when you have the 30X focused and you turn to the 100X, does the image remain in focus? (This is a very nice feature if you can afford it, although my experts tell me that the parfocal systems often don't work.)

√ Look at features not listed here—special packaging, dissecting tools, and the like. Remember that you are paying for these features, even if they are ostensibly included in the price. Are they worth it?

√ Look at the documentation that accompanies the microscope. Do the experiments seem interesting and inviting? Are replacement parts available? Can the system be improved and expanded? Are guarantees and warrantees included?

√ If possible, avoid purchasing a microscope that you haven't seen. While comparing microscopes, bring a homemade slide to the nearest science store and view it through several different scopes. Needless to say, manufacturers and their representatives want to show you their best and most dramatic slides at the time of purchase.

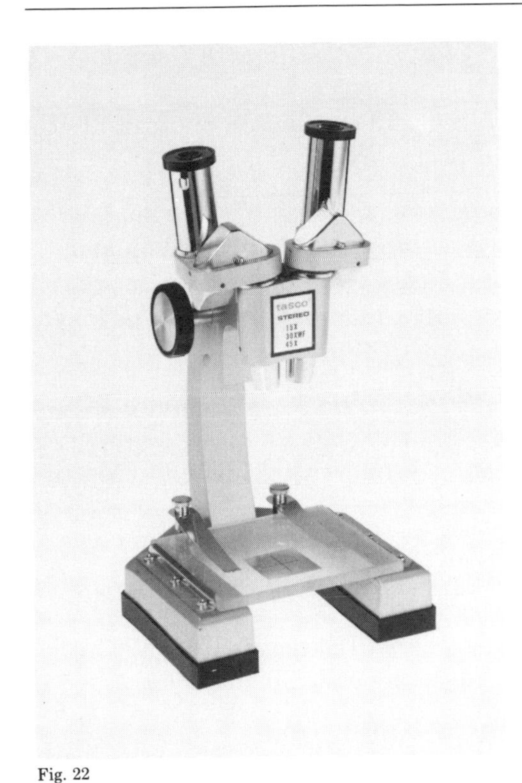

Fig. 22

Fig. 23

Fig. 22. Because of the duplicated optics, stereo microscopes tend to be more expensive than compound microscopes. A beginner stereoscopic device, like this recommended instrument from Tasco, retails for approximately $200.

Fig. 23. The MONOCULAR from Jason Empire brings objects 8 times closer with amazing clarity and optical brilliance. It comes complete with a deluxe leather carrying case, and retails for approximately $100.

IN PRINT

Anderson, Lucia, *The Smallest Life Around* (Crown, 1978).
The author discusses fascinating characteristics of microbes young biologists can see for themselves.

Beeler, Nelson F., and Franklyn M. Branley, *Experiments with a Microscope* (Crowell, 1957).
Although this book should surely be updated, it still provides lots of good advice for parents and older children about setting up interesting experiments for themselves or younger siblings.

Grave, Eric V., *Discover the Invisible: A Naturalist's Guide to Using the Microscope* (Prentice-Hall, 1984).

A well-done travelog to microscopy with useful sections on photographing through the microscope. For advanced beginners.

Grillone, Lisa, and Joseph Gennaro, *Small Worlds Close Up* (Crown, 1978).

Full-size black-and-white photos, taken with the most advanced optical equipment, are enjoyed by children of all ages. Note that what you see under your compound microscope bears little relationship to photos taken by electron microscopists.

Headstrom, Birger Richard, *Adventures with a Hand Lens* (Lippincott, 1962).

With these suggestions and clear directions, children are invited to see an amazing world that was there all the time.

———, *Adventures with a Microscope* (Dover, 1977).

Clear directions and lots of good advice for the project-hungry amateur.

Kaveler, Lucy, *Green Magic: Algae Rediscovered* (Harper & Row, 1983).

The characteristics and function of algae, a one-celled plant, is described in detail for advanced readers.

National Geographic Society, *Hidden Worlds* (National Geographic Society, 1981).

Describes how contemporary imaging techniques—x-ray, ultraviolet, electron microscopy, astrophotography—are used to make the heretofore unseen, visible.

Selsam, Millicent, *Greg's Microscope* (Harper & Row, 1963).

As Greg looks at typical household items with his new instrument, your child can do the same. For the youngest microscopists.

Simon, Seymour, *Exploring with a Microscope* (Random House, 1969).

More excellent activities for junior scientists.

13 Biology: Experimental Thinking

Any time you redesign an environment to see "what would happen if," you are experimenting. Experimentation and unobtrusive observation (the kind of biology described in Chapter 11) are not necessarily competing methods. One could, for example, note that certain lichen prefer moist shade to dry shade: Move a rock with the dry lichen to a spot where moisture-loving snake grass grows; leave another rock with lichen in the dry spot; and then check growth patterns. In this instance the idea for the experiment came from passive observation, while the experiment itself involved some intervention. This experiment required no "science things."

For other experiments, however, reliable equipment is needed; a thermometer, for instance, is a useful tool for a child studying the effect of temperature on the growth of molds. Model experiments, often found in books, are also an important resource. They give amateurs a sense of what an experiment looks and feels like, and some of the satisfactions involved in this kind of activity.

HANDS-ON

For Studying Plants

Gifts for young botanists need not always come as prepackaged sets. Flower bulbs to be forced in winter, gardening tools, or seeds for favorite foods are often received with great excitement. If, however, you are in the market for a more formal botany kit, here's what's available.

Fig. 24

Fig. 25

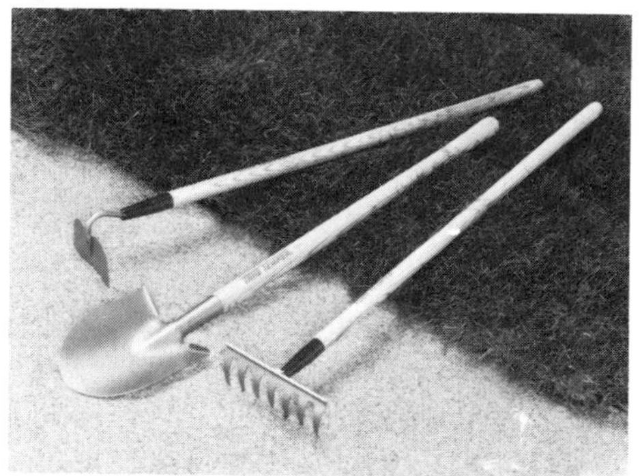
Fig. 26

MUSHROOM FARM-IN-A-BOX, approximately $11 from Edmund or Nasco, provides you with everything you need to grow several crops of mushrooms over a three-month period. Add water, take careful notes, and harvest your first crop in thirty days.

Fig. 24. A DWARF TREE KIT from Uncle Milton Industries grows nine Bonsai trees and includes soil mix, food, and a ninety-page booklet. Available from Edmund and other science stores for $13.

Hydroponics sets which teach children how plants are grown in enriched water are a "growing fad."

Fig. 25. This HYDRO-GREENHOUSE from Uncle Milton Industries uses no soil; it comes with a pump which recycles water, plant nutrients, and air. The 10″ × 10″ unit produces as much as would 10 square feet of ground space, and sells for about $40 from science supply houses and hobby shops.

The HYDROPONICS "NURSERY UNIT" from Carolina Biological (order #66-6860) includes a styrene tray, seven self-watering inserts with nine compartments each, nutrient reservoir, plant food, plant germination covers, markers, and an instruction booklet; available for $22.50. This one is well designed for a school or club.

For the more traditional indoor gardener, the 24″ × 24″ × 18″ CRYSTALLITE INDOOR GREENHOUSE from Nasco reduces time to maturity by 40%. Fold-down construction and automatic watering with rot-proof fiberglass wicks are a plus. Basic unit is $30; approximately $50 with fluorescent double tube fixture.

Fig. 26. Real garden tools, the perfect size and weight for young children, are available from Toys to Grow On. The shovel, rake, and hoe, made of heavy-gauge steel with a non-rusting bronze finish and securely mounted on 36″ ash handles, sell for about $30 for the set.

For Studying Animals

Studying the behavior of animals in a controlled environment is fascinating. (For guidelines on the responsible treatment of animals used in experimentation, see pages 66–67.) In nature it is difficult to see ants actually tunneling underground, or butterflies forming a chrysalis, but with the clear animal housing sold commercially the wonders of life become readily apparent. There are two major problems, however, with "animal farms" which should be recognized.

First, the clear housing sold by manufacturers creates an environment so foreign to the species in question that the observed behavior may not be "normal" animal behavior. Second, the animal "farms" one purchases do not include the species in question, but rather a coupon to be redeemed for frog eggs, ants, or whatever. Most kids see this as inconvenient.

But inconvenience is only half the story. Biologists and government officials agree that animals or plants from one area must not be unnaturally introduced into another area. They can upset a delicate ecology and spread disease. Thus, when ants for the ant houses, for instance, are shipped they do not include a queen ant. Without a queen the workers cannot reproduce, and if they don't reproduce, there's no worry about introducing non-native animals into the area. On the other hand, if the ants can't reproduce, they die out, generally within six to eight weeks. No amount of food or loving care from humans can keep them going.

There are three possible solutions to this problem. One could decide to buy a commercial "animal farm," send in the coupons, watch the animals grow, then watch them die. Most kids find this fairly discouraging. Another possibility is to buy the commercial housing, fill it with species you collect from your neighborhood, and eventually, after you've studied them, return the animals to nature. (To find your own queen ant, look for a relatively small anthill, and with one big shovelful, heap it into a cardboard box. Using a strainer or screen, sift through the dirt looking specifically for the largest insect; there is only one queen per hill and she is significantly bigger than the worker ants.) The third alternative is to make your own housing and find your own animals for study. There are a number of excellent books that pro-

Fig. 27

Fig. 28

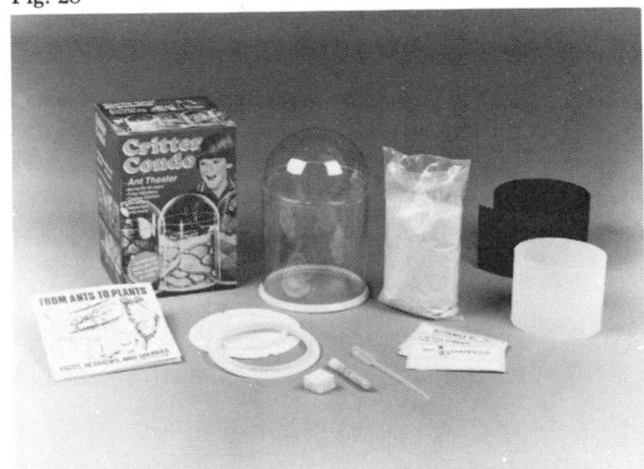

Fig. 29

vide directions for undertaking this and other similar projects (see, for instance, Seymour Simon's *Pets in a Jar: Collecting and Caring for Small Wild Animals,* Viking, 1975).

In any case, with any "animal farm" gift, include a book about the species in question so that children have some help in observing and understanding what they see (for age-appropriate recommendations, see pages 171–175).

In the two ant farms currently available, one produced by Natural Science Industries (fig. 27) and the other from Uncle Milton Industries (fig. 28), the ants and dirt are sandwiched between two transparent plastic sheets. These kits sell for $10 to $17 each, depending on the size and design of the housing.

Ant experiments, not discussed in the prepackaged directions, are altogether possible. To suggest a few: The child could pack the housing with different kinds of sand and soil to see which the ants prefer. A variety of ants could be collected and studied. Children might keep notes on eating preferences and habits, and longevity patterns could be examined.

Butterfly gardens and frog hatcheries, which have the same assets and liabilities as the ant farms, also require the child to redeem a coupon for the requisite larvae and frog eggs.

Fig. 29. The **CRITTER CONDO,** complete with certificates for butterfly larvae, cacti, ants, and frogs, is available from Nasco (among other places) for $12.50.

Fish, gerbils, guinea pigs, mice, dogs, cats, and so on give children a sense of what it means to care for an animal as it matures. Martin's Aquarium, 101 Old York Rd., Rt. 611, Jenkintown, PA 19046 (also of Cherry Hill, NJ), distributes a free catalog of pet supplies, including small-animal cages, aquaria, terraria, and birdhouses. Bargain hunters should check these prices before buying elsewhere.

Fig. 30. This attractive **BUTTERFLY GARDEN,** available from Nasco and in some toy stores, retails for $13. The **FROG HATCHERY,** also from Nasco, contains an electrically lighted observatory with microscope viewer, food, book-

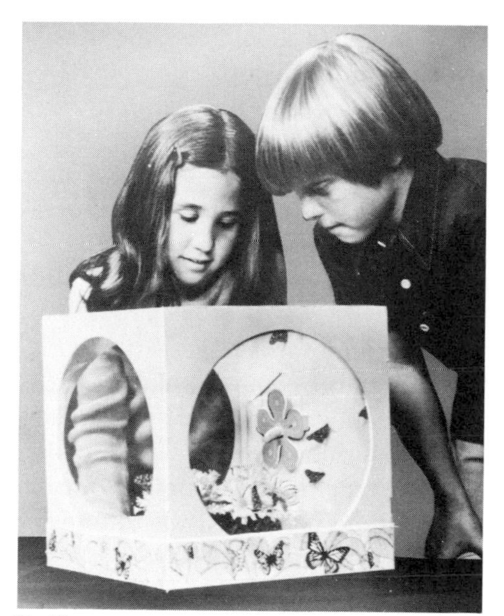

Fig. 30

let of instructions, and coupon for a supply of frog eggs for $16.50. The frog eggs are also sold separately for $3. *Warning: Do not free mature animals in an area where they are not native.*

The four-egg **CHICK INCUBATOR** from Marsh, available through Toys to Grow On and science supply houses for $23, unlike most small incubators really works. For a farm child, a school, or city people able to find suitable homes of chickens or ducks, this is a good gift, although it requires adult help. Several excellent books on the development of chicks are listed below.

Fig. 31. Books and toys needn't be thought of as distinctly different items. *Beastly Neighbors* by Mollie Rights (Little Brown, 1981), for instance, provides excellent directions for making terraria, flower presses, ant farms, and more.

A worm farm from Carolina Biological contains a styrofoam box with worm bedding, worm food, and an instruction pamphlet, for $7.50. For another $2.25 this same kit comes equipped with red worms. (Nightcrawlers are slightly more expensive.) This product enables children to find cocoons, watch as worms grow, and make an important contribution to the family garden.

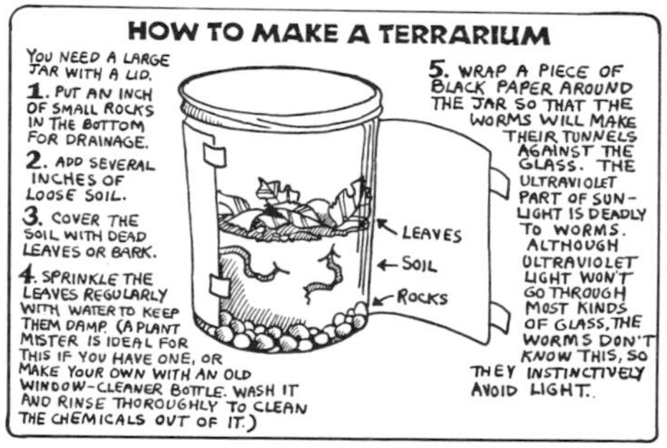

Fig. 31

Dissection is the most intrusive biological exploration. There is significant discussion about whether children should, in fact, dissect animals: Is what they learn worth the sacrifice of an animal? What values do they take away from the experience?

Biologists claim that dissection is an important part of their science. Dissected organisms simply look different than photographs or drawings. An intimate understanding of an animal, both inside and out, teaches an appreciation of all living systems. And finally, there is a hands-on aspect to dissection that is important for people who wish to understand what scientists do.

If you and your children decide that dissection is an activity you sanction, certain rules should be made clear. Children should never be encouraged to

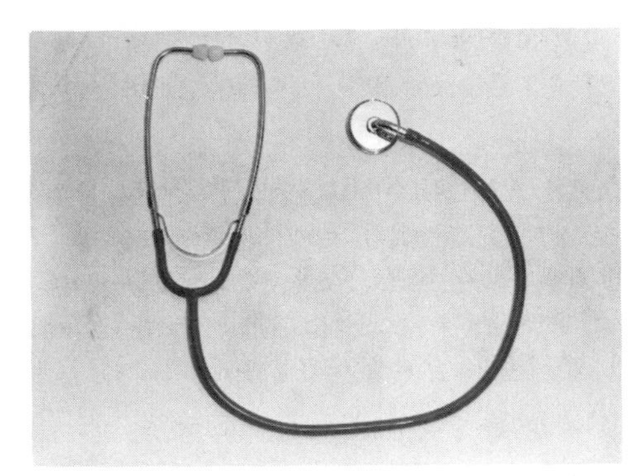
Fig. 32

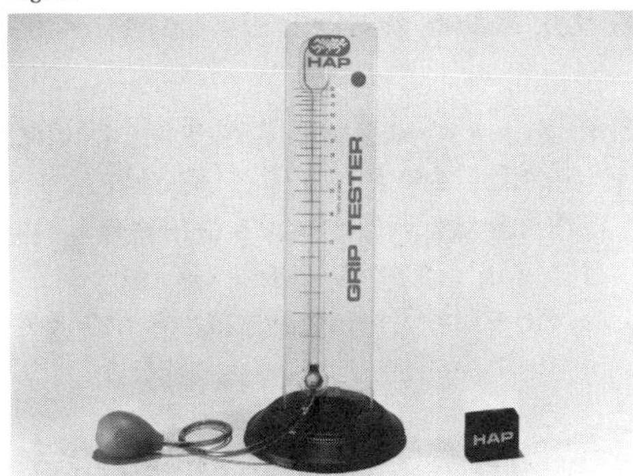

Fig. 33

find animals, then kill and dissect them. Live animals should be viewed as part of a complex ecosystem that we have no right to disturb. *Moreover, even is a child finds a dead animal, I do not recommend dissecting it; it may well be diseased.*

For people interested in teaching their children about the insides of living systems, the best alternative is to dissect animals you are about to eat, or have just eaten. A fish carcass, a clam, a chicken, or even an egg can nourish the mind as well as the body. Dissecting plants is also a good way to develop technique and learn about a living system.

If you feel competent to lead a child through an informed dissection with a preserved specimen, purchase these from a reputable company like Nasco. Their catalog is free to parents, whereas most catalogs which list preserved animals cost $6 to $8, and they claim that their formaldehyde levels are extremely low. Even so, remember that formaldehyde (both the liquid solution and the fumes) is a dangerous chemical and should be treated with utmost care and caution. All science supply houses carry dissection kits.

Note: Please do *not* let children believe that just because they can cut up an animal, or catch and mount butterflies, they are doing science. Many science fair entrants come away disappointed and sometimes ashamed when judges point out that they had no reason for cutting and collecting.

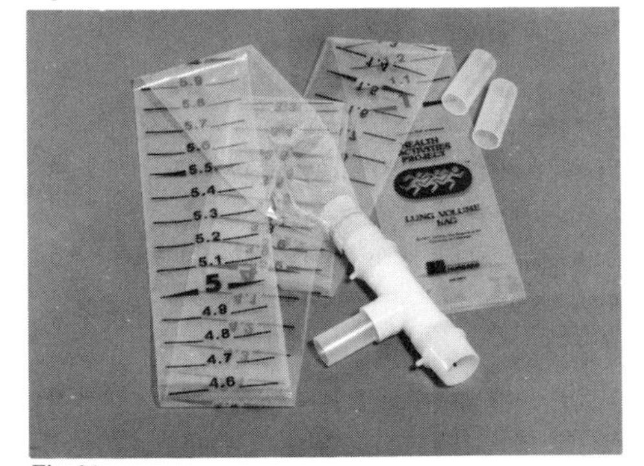

Fig. 34

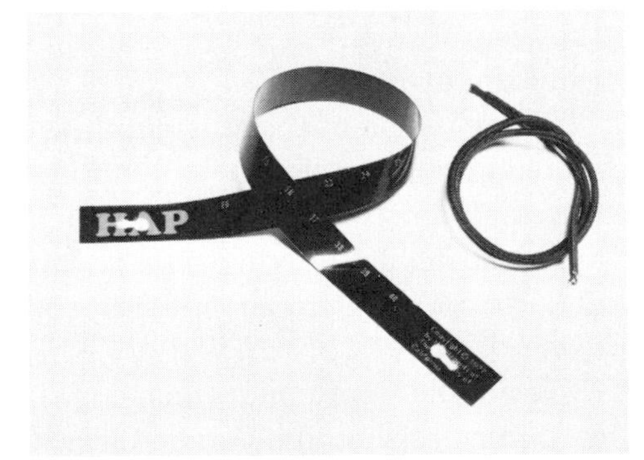

Fig. 35

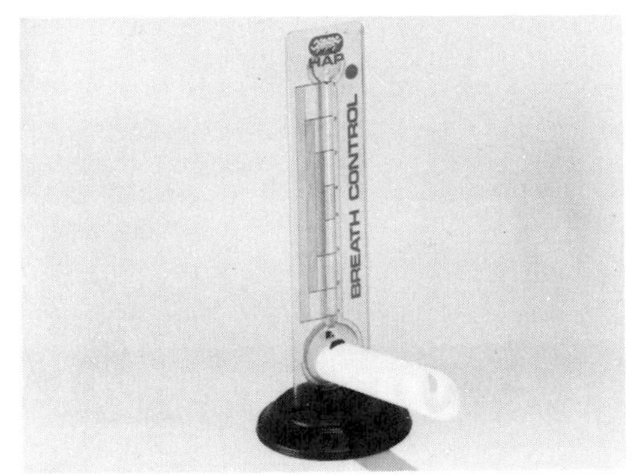

Fig. 36

For Studying Ourselves

For All Ages

Describing oneself in scientific terms, and then looking for changes, is interesting for children of all ages. At regular intervals, mark children's height and weight gain in a place they can see. Teach children how to count pulse beats; note changes in pulse after exercise.

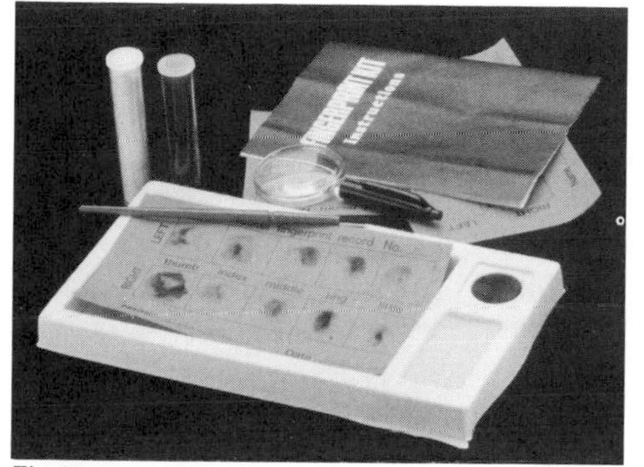

Fig. 37

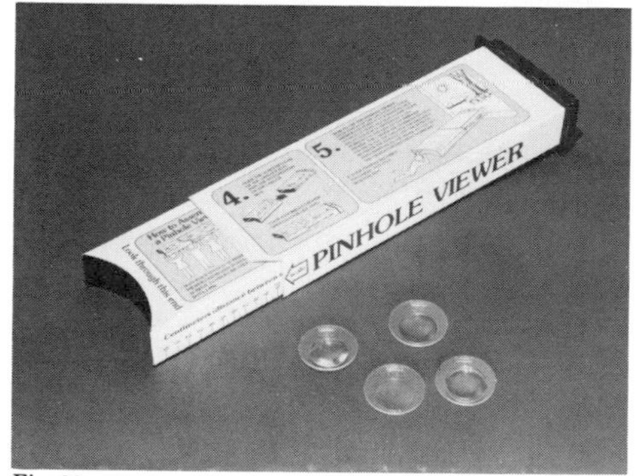

Fig. 38

Use a real stethoscope to listen to heartbeats. Edmund has one in their SCIENTIFICS line ($11) and Hubbard sells a reliable plastic model through museum shops ($4).

Although I don't recommend giving children a thermometer filled with mercury for "play," a **LIQUID CRYSTAL THERMOMETER,** which looks and feels like a headband, is available through Hubbard for $2. Readings can be taken when one wakes in the morning, after exercise, or in the evening bath.

A **GRIP TESTER** ($12), **LIMBER GAUGE** ($12), and **BREATH CONTROL DEVICE** ($10) are also available through Hubbard. Using these devices, the effects of food, exercise, allergens, and time of day can be noted.

Figs. 32–36. Pictured here are the stethoscope, grip tester, lung volume bag, liquid crystal thermometer, and breath control device from Hubbard's series of health-related devices.

Fig. 37. Fingerprint kits, available from both Natural Science Industries (NSI) and E.D.I. ($4 to $10), are wonderful gifts. While children under eight work hard at making and lifting prints, their older siblings can practice observation and classification skills.

Fig. 38. The **PINHOLE VIEWER** from Hubbard provides an excellent demonstration of how the eye works. It can also double as a pinhole camera.

A pad of **FOOD SCORECARDS** ($1) from Play 'n Peace help children evaluate their nutritional intake.

For Young Children

Pretending toys, for instance the Fisher-Price **DOCTOR KIT,** stimulate important conversations about the placement and function of body parts, and the role of physician and patient in keeping well.

Puzzles (and drawings) help children think concretely about what goes where. A four-layer wooden SEE INSIDE PUZZLE (girl and boy versions available), which shows the skeleton, organs, veins and arteries, and skin, is available from Toys To Grow On for $17.

For Older Children

As a way of knowing what goes on inside a living organism, biological models have two distinct advantages over books: first, they give an important three-dimensional view of the organs, and second, in putting them together the child must think about what goes where. The suggested age (twelve and above) for the following kits has more to do with dexterity than comprehensibility. Thus, if you are looking for a good family project, one of these may be it.

Figs. 39–40. Lindberg produces a series of anatomically correct, comparatively inexpensive ($7 to $10 each) models of a Beating Heart, Brain/Skull, Human Skeleton, Visible Man, Visible Woman, Nose/Mouth, Human Ear, Human Eye, and Human Skull. Available from toy stores, hobby shops, and science catalog houses.

Fig. 41. A full-size HUMAN SKELETON, poseable, scientifically accurate, and wonderfully amusing, is assembled from die-cut cardboard sheets, with specially provided pins. Available for $49.95 from science supply houses, compre-

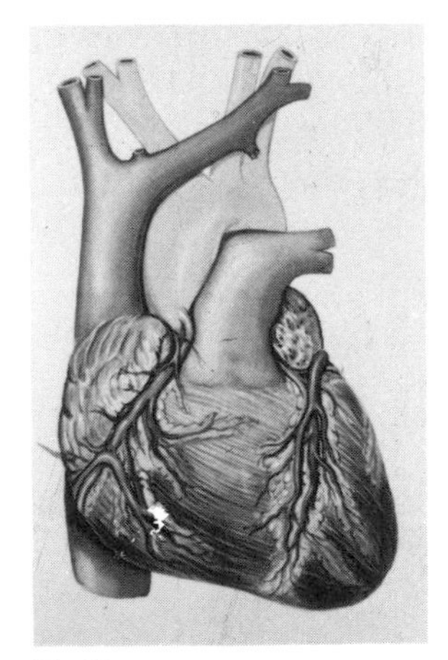

Fig. 39

Fig. 40

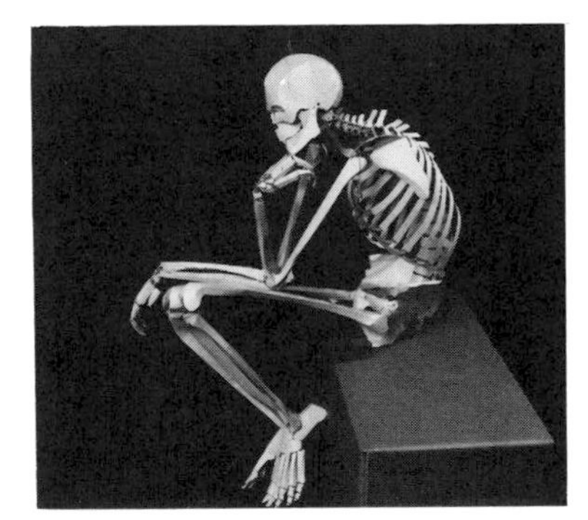

Fig. 41

hensive bookstores, or direct from Albion Import Export Co., Coolidge Bank Bldg., 65 Main Street, Watertown, MA 02172 (plus $3 postage and handling).

Fig. 42

Fig. 42. Wooden models are stylized and artistic representations of nature's own. Although these figures are not particularly good for teaching animal anatomy, they do give a general sense of the species' shape, and serve as a tribute to the life-form. Waco, a Japanese importer, is U.S. distributor of an exquisite series of insects, dinosaurs, and birds which range in price from $7 to $27; available mainly through better toy stores, museum and gift shops.

A game, **SUPER SANDWICH,** from Teaching Concepts, Inc., compares favorably with any board game around. Players collect enough vitamins, minerals, and calories to meet the recommended daily requirement. Suddenly milk is seen as a great bargain for 25 cents; while rolling the dice, small children call out for the right number to land on spinach, an excellent source of vitamin A. Too many calories and one is sent to the gym to work them off, thus losing a turn. Although small liberties were taken in adjusting the number of calories found in certain foods, the basic principle is scientifically correct. In short, the game is more than palatable. Recommended for children age ten and older. Sells for $16 at better toy stores, museum shops, and science supply companies.

IN PRINT

For All Ages

Armour, Richard, *Insects All Around* (McGraw-Hill, 1981).
Funny but factual verse about nineteen insects, including the flea and the grasshopper.

Bendick, Jeanne, *How Heredity Works: Why Living Things Are as They Are* (Parents, 1975).
Genetics, reproduction, and speciation as related to plants, animals, and humans are explained in language non-scientists can readily understand.

Miller, Jonathan, *The Facts of Life* (Viking, 1984).
A three-dimensional explanation of reproduction, from conception to birth. This unusual pop-up book is outstanding.

———, *The Human Body* (Viking, 1983).
The heart beats, the lungs breathe, and muscles contract in this pop-up book which accurately reveals the mysteries of the human body.

Selsam, Millicent, *Play with Plants* (Morrow, 1978).
This revised edition of a classic experiment book is accessible, inviting, and satisfying.

Stern, Sara Bonnett, *Making Babies: An Open Family Book for Parents and Children Together* (Walker, 1974; paperback, 1983).
Photos, large type, and simple text tell the story to children; a fuller explanation is provided for parents in small type.

For Young Children

Although even young children can appreciate the ant farms and stethoscopes recommended in this chapter, the books to accompany such gifts should be chosen with their interests and reading abilities in mind. These selections go well with the toys and equipment listed earlier.

Berger, Melvin, *Why I Cough, Sneeze, Shiver, Hiccup and Yawn* (Crowell, 1983).
With an artfully light touch, Berger answers questions about reflex reactions.

Cobb, Vicki, *Lots of Rot* (Lippincott, 1981).
Lively information and experiments about a most natural process.

Cole, Joanna, *How You Were Born* (Morrow, 1984).
A four-page parent guide precedes a cogent explanation of human conception and birth. With numerous photos showing embryonic and fetal development, labor, and newborn babies, this book is remarkable for its clarity.

Henley, Karyn, *Hatch!* (Carolrhoda, 1980).
Describes the nesting habits of egg-layers: whale sharks, Siamese fighting fish, grasshoppers, ostriches, platypuses, and others.

McCauley, Jane R., *Baby Birds and How They Grow* (National Geographic Society, 1983).
Baby birds, their parents, nests, and early care are examined in this simple, large-format text with exceptional color photographs.

McGovern, Ann, *Shark Lady* (Four Winds, 1979).
Dr. Eugenie Clark became interested in fish at age nine during regular visits to the New York Aquarium. A good biography.

Oechsli, Helen and Kelly, *In My Garden: A Child's Gardening Book* (Macmillan, 1985).
The perfect book to accompany a set of gardening tools, this colorful and comprehensive guide to the ins and outs of vegetable care includes hints for all ages. An attractive book that makes harvest time all that it should be.

Selsam, Millicent, *Where Do They Go: Insects in Winter* (Four Winds, 1982).
Explains how bees huddle, butterflies migrate, and more.

Showers, Paul, *What Happens to a Hamburger* (Crowell, 1985).
This newly revised edition of the digestion classic includes experiments and explanations to convince young readers that food gives energy and builds strong bones and muscles.

———, *You Can't Make a Move Without Your Muscles* (Crowell, 1982).
A casual introduction to human movement, illustrated with cartoon-like illustrations.

For Grade-Schoolers

Allison, Linda, *Blood and Guts: A Working Guide to Your Own Insides* (Little, Brown, 1976).
Provides explanations and lots of experiments to help users better understand the workings of the human body.

Epstein, Sam and Beryl, *Dr. Beaumont and the Man with the Hole in His Stomach* (Coward, 1978).
Biography of the physician and his patients—an amazing tale of creative experimentation.

———, *Secret in a Sealed Bottle* (Coward, 1979).
How Lazzaro Spallanzani disproved the theory of the spontaneous generation of microbes.

Jacobs, Francine, *Breakthrough—The True Story of Penicillin* (Dodd, 1985).
A thought-provoking tale of the ways social circumstance, luck, and science combine to create what we call "medical advances."

Johnson, Sylvia A., *Inside an Egg* (Lerner, 1982).
Amazing photographs present an informative, unusual look at bird growth.

Kelves, Bettyann, *Thinking Gorillas: Testing and Teaching the Greatest Ape* (Dutton, 1980).
Stories of individual gorillas, how they were studied and tested.

Lauber, Patricia, *What's Hatching out of That Egg?* (Crown, 1979).
All about eggs and the animals that hatch from them, including ostrich, python, and bullfrog.

Overbeck, Cynthia, *Ants* (Lerner, 1982).
About ants that build nests, travel, and invade other nests.

Robinson, Marlene M., *Who Knows This Nose?* (Dodd, Mead, 1983).
Nasal characteristics and function are featured in this entertaining book which asks children to identify animals from close-up photos of their noses.

Sattler, Helen R., *Fish Facts and Bird Brains: Animal Intelligence* (Lodestar, 1984).
General information about animal intelligence with suggestions on ways to test the abilities of your own pet.

Selsam, Millicent, *Eat the Fruit, Plant the Seed* (Morrow, 1980).
Directions for growing mango, avocado, papaya, citrus, pomegranate, and kiwi.

For Middle-Schoolers

Bornstein, Sandy and Jerry, *New Frontiers in Genetics* (Messner, 1984).
The science of genetic research is related to social, moral, and political issues in this thought-provoking volume.

Flanagan, Geraldine L., *Window into a Nest* (Houghton Mifflin, 1976).
The development of a chick embryo from fertilization until it hatches.

Gilbert, Sara, *Using Your Head: The Many Ways of Being Smart* (Macmillan, 1984).

A reassuring and scientifically accurate discussion of intelligence.

Haines, Gail K., *Brain Power: Understanding Human Intelligence* (Watts, 1979).

The functions and physiology of the brain, as well as possibilities for brain alterations are discussed here.

Hopf, Alice L., *Nature's Pretenders* (Putnam, 1979).

An intriguing look at ways plants and animals defend themselves from natural enemies.

Kapit, Wynn, and Lawrence M. Elson, *The Anatomy Coloring Book* (Harper & Row, 1977).

For a kid determined to learn anatomy, this sophisticated coloring book and explanatory material will do the trick. A set of colored pencils, and a note about the difficulty of the task, should accompany the gift.

Johnson, Sylvia, *Mushrooms* (Lerner, 1982).

How fungi grow and reproduce—a photo essay.

Peavy, Linda, and Ursula Smith, *Food, Nutrition and You* (Scribner, 1982).

This anecdotal overview of nutrition includes information on eating disorders, sports and nutrition, food additives, vitamins, and fast foods.

Prime, C. T., and Aaron E. Klein, *Seedlings and Soil: Botany for Young Experimenters* (Doubleday, 1973).

Young people investigate structure, function, and behavior of plants. Practical and inexpensive advice for the home experimenter.

Wilson, Ron, *How the Body Works* (Larousse, 1978).

Emphasis on how the system works together, with a nod to the modern hospital and the history of medicine.

14 Digging Earth Science

Geology for kids is popularly associated with "rock hunting." Children like to collect and name things, and rocks, because they are abundant and easy to find, are perfect for young pack rats. Geologists generally agree with this commonsense notion that rock collecting is great fun, but add the following important caveats:

- What scientifically naive people think of as rocks, geologists call minerals, rocks, and fossils. Rocks, in the technical sense, are composed of irregular portions of minerals, and are often hard to identify, even for experts. Minerals, on the other hand, are made up of regular, repeating patterns of atoms which, with a few physical tests, can be discriminated one from the other. Moreover, the scheme used to classify them is much like the hierarchical, binomial scheme used to classify plants and animals. Fossils are imprinted remnants of a life long past. For young collectors, fossil hunting is extremely rewarding; fossils bear a close relationship to shapes children recognize, and in areas rich in specimens, kids are bound to find lots of interesting remains.
- Geology is fundamentally a study of time, a study of the history of the earth. Because the time span described is so huge, it is hard to appreci-

ate. And yet, to simply talk of specimens as old (for instance, a 10-million-year-old rock is extremely young) is to ignore the comparative sense of age that gives meaning to all that geologists do. For this reason, children should be encouraged to discuss not only the physical traits of specimens, but also their history.

HANDS-ON

For Collecting and Identifying

Collecting and identifying specimens are activities as interrelated as pedaling and steering a bicycle. Although you could do them separately, to get anywhere both are necessary. Kids learn to collect and identify in two ways, either by finding specimens and using a field guide to identify them, or by using a mineral identification kit. In either case, the same physical tests are used to differentiate specimens from one another:

- The way the mineral breaks
- Its hardness
- Its approximate specific gravity: light, normal, or heavy
- Properties depending on light—color, luster, transparency
- Magnetic properties—will it attract a magnet?
- Chemical properties—will it fizz with dilute hydrochloric acid (HCl)?

Note: Even dilute hydrochloric acid is a potentially harmful acid. Do not let anyone drink it, and if you spill some on you, wash the area immediately with lots of cold water.

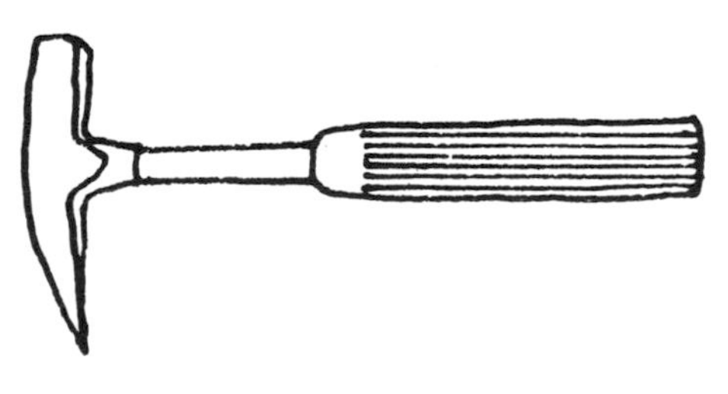

Fig. 43

Fig. 43. Two popular types of geologist's *hammers* are available in hardware stores or science supply houses. The geologist's *pick,* pictured here, is particularly good for dislodging specimens; the geologist's *chisel* is generally used by fossil hunters or collectors working in soft sedimentary rock. *Warning:* Do not use one hammer to strike the other, and do not use the pick for striking.

Useful Identification Items for the Field Collector

A needle or nail for testing hardness

A magnet

A dropper bottle of muriatic acid (1/10 Normal hydrochloric acid)

A knapsack to carry supplies and bring back specimens

A notebook to keep track of where and when the specimen was found

Paper with which to wrap individual specimens

A pocketknife which can be used to test hardness or loosen dirt

A 6 to 10X hand lens

Goggles for obvious safety reasons

A geologist's hammer

Recommended Guides

A field guide or a book about identification is also a must. The classic text is C. S. Hurlbut and E. S. Dana's *Minerals and How to Study Them* (Wiley, 1949). Though intended for adults, the information given here is so well organized that any parent or older child in need of information will find it useful.

For those determined *rock* hunters, Russell P. MacFall's *Rock Hunter's Guide* (Crowell, 1980) gives information on identifying, collecting, and displaying your finds.

Other guides are intended mainly for children.

Ritter, Rhoda, *Rocks and Fossils* (Watts, 1977).
A good elementary description of geological features—caves, shores, and meteorites as well as sedimentary, igneous, and metamorphic rock.

Shedenhelm, W. R. C., *The Young Rockhound's Handbook* (Putnam, 1978).
Filled with practical advice for new and experienced collectors, this one can be used by independent older kids or young children with parental

Tips for Collectors*

- Always note where a specimen was found in a field notebook. Identification can wait, but location should be recorded immediately.
- Ask permission to collect rocks, minerals, or fossils on private property. Owners are thankful for the courtesy, and you will not be trespassing.
- Do not collect specimens in national parks or monuments, or in state parks where it is illegal. Similar specimens commonly crop out on nearby land and can be collected there.
- Practice identification. Kids may quickly learn to name most minerals in your area. Then try characterizing the rocks and minerals used to build houses or offices, or those used in cemeteries. Visit museums and dealers. A pamphlet entitled *Mineral, Fossil and Rock Exhibits & Where to See Them* from the American Geological Institute is a must for museum-goers.
- Start or join a geology club. Subscribe to a collector's magazine.
- Do not collect a specimen from each state or country you visit, this kind of collection has no scientific meaning. Natural phenomena know no political boundaries.

help. Especially recommended is the section "Sixty Common Minerals and How to Tell Them Apart."

Identification Kits

The other approach to identification is offered by manufacturers of rock and mineral kits who provide specimens that children can memorize, handle, and match with their own finds. At their best, mineral kits give kids hands-on experience with the items in question and expose them to a greater variety of specimens than they are likely to find in their own neighborhood. Be advised, however, that the quality of samples varies, and that the included minerals may not match specimens in your area.

* Information adapted largely from U.S. Government Printing Office brochure, "Collecting Rocks," by Rachel M. Barker of the U.S. Geological Survey.

Fig. 44

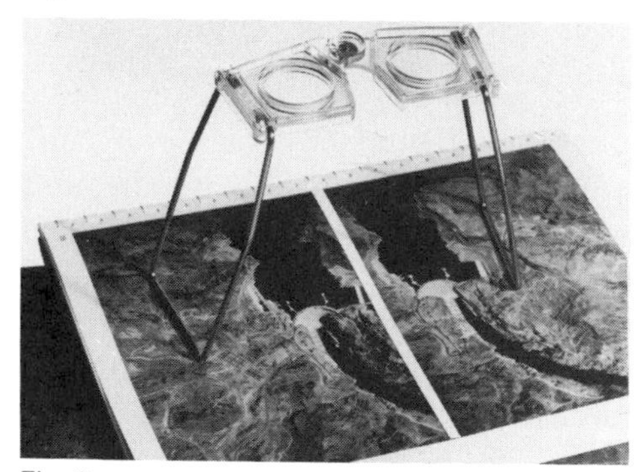

Fig. 45

Edmund has a good 36-specimen kit for $12, and a 100-specimen set for $30; Nasco sells a PHYSICAL PROPERTIES OF MINERALS SET of 26 specimens for $15.40, with directions on how to test what you find. For school-size collections, get the Ward's or Scott Resources catalog.

Fig. 44. NATURIFICS, a line from Edmund, includes geological samples children can handle and study, plus a fairly sophisticated description of the object and its scientific import. Of particular interest to the earth science fan are: Pumice Stone/Rocks that Float ($3); Lodestone/Nature's Magnet ($3); Tektite/Rocks from the Sky ($3.50); and The Coral Reef/Underwater Cities ($4).

A subscription to the Mineral of the Month Club (13057 California St., Yucaipa, CA 92399) can be a wonderful gift for serious rockhounds. Each month participants receive specimens, good scientific documentation, ideas for experiments, and a newsletter in which hobbyists list items they wish to buy, sell, or trade. Although the twelve-month subscription fee is $50, one can sign up for a trial period of three months for $15. *Note:* On occasion, descriptive prose is laced with Christian doctrine.

Fig. 45. Hubbard Scientific takes a novel approach to the problem of identification with stereoscopic books, one on ROCKS, MINERALS AND GEMS and another on FOSSILS, which provide dramatic 3-D images of the specimens displayed. The books are about $7 each; $3.25 for the 3-D glasses.

Fig. 46. This rock polishing kit, $26 from N.S.I., includes a motorized tumbler and abrasive material; you supply the electricity. More expensive versions of these sets (up to $60) include larger barrels and bigger motors. Rock polishing and jewelry making can lead to a more scientific interest in geology. A fun hobby for young and old.

Fig. 46

INTRODUCING PALEONTOLOGY

Using fossil remains as clues, paleontologists study the earth's history. Because fossils are embedded in rocks, much of the advice for rock collectors is also applicable to fossil hunters. But fossil identification is based on biological classification as well as physical geological tests; certain plants and animals are characteristic of specific eras and environments. To provide information of this sort, use one of the following books.

Aliki, *Fossils Tell of Long Ago* (Crowell, 1972).
Describes how fossils are formed, what they tell about the past, and where they are found. Aliki has a wonderful light touch. Especially recommended for young children.

Keen, Martin L., *Hunting Fossils* (Messner, 1970).
A lively discussion of fossils and how to find them, appropriate for gradeschoolers.

Lambert, Mark, *Fossils* (Arco, 1978).
A comprehensive guide to the hunting and meaning of fossil remains, intended for the advanced reader.

Zim, Herbert, Frank Rhodes, and Paul Schaffer, *Fossils* (Western, 1962).
An inexpensive ($3) identification guide, useful to enthusiasts of all ages.

Fossils and Fossil Kits

Fig. 47. 20 SELECTED PLASTIC FOSSILS with an instruction manual by geologist Daniel Jones is sold through Hubbard Scientific for $4.95. Although the manual is written for advanced readers, the concepts of classification and the information provided are of interest to younger children too.

Fig. 48. FOSSIL HUNT from Uncle Milton Industries ($20) invites children to sort through a box of volcanic sand and crushed rock for more than twenty actual fossil specimens. Set includes magnifier, brushes, and 160-page guide, and is available from Edmund and other houses.

A STARTER FOSSIL KIT, available for $8.50 from Schoolmasters, includes ten different representative fossils of the Pennsylvania Period of the Paleozoic Era, and a book, *How to Build a Fossil Collection,* which explains collecting and identification procedures to the novice.

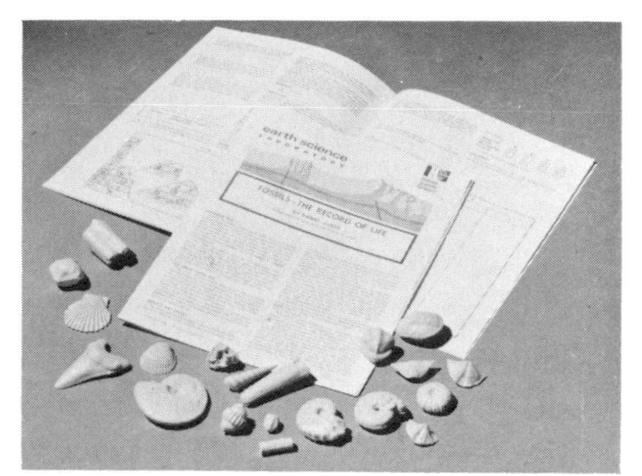

Fig. 47

Fig. 48

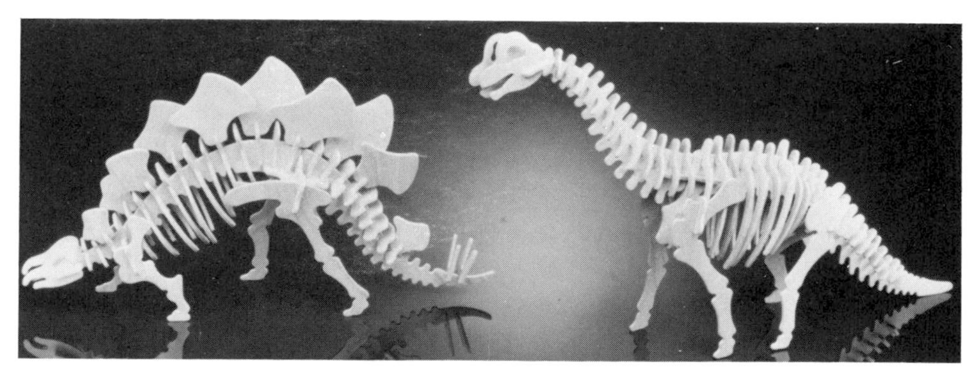

Fig. 49

Models of Ancient Beasts

Fossils are imprints left by animals and plants long gone. For children, the most impressive of those ancient life-forms are vertebrates—dinosaurs, archaic birds, and wooly mammoths. Reproductions of these early species come in two forms. There are "fleshed out" replicas (the best of these are taken from British Museum molds and are available through Edmund or The Nature Company: a pteranodon for $3, a triceratops for $5, and a wooly mammoth for $6) which children collect on a shelf or use for dramatic play. There are also skeletal models which give kids some sense of the animal's bone structure. (Attractive wooden models distributed by SMALL WORLD TOYS sell for $6–$26, while plastic snap-together skeletal models from Battat (fig. 49) can be had for $5.) Building the models is fun for children or their parents, but the skeletal versions all fall apart easily and thus are no good for dramatic play. Natural-history museum shops generally carry both types. For suggestions on how to stimulate questions using these models see pages 103–104.

IN PRINT

Maps

Because geologists are concerned with landforms and locations, maps are invaluable resources. For the amateur or advanced collector, there is no better source of information than *Maps and Geological Publications of the United States: A Layman's Guide,* compiled by William R. Pampe, available for $3 from the American Geological Institute, 4220 King St., Alexandria, VA 22302. This book is a state-by-state listing of general information pamphlets; maps; location-specific rock, mineral, and fossil data; and information on various landforms—caves, volcanoes, water resources, etc. It is a spectacular array of information.

The National Cartographic Information Center of the U.S. Geological Survey provides the public with aeronautical and nautical charts, city maps, extraterrestrial maps, forest maps, geologic maps, highway maps, land-use maps, river surveys and dam site maps, soil maps and topographic maps, cartographic educational materials, atlases, gazetteers, and aerial and space imagery. Contact the National Cartographic Information Center at the U.S. Geological Survey, 907 National Center, Reston, VA 22092.

EROS (Earth Resources Observation Systems) Data Center makes LANDSAT satellite imagery, National Aeronautics and Space Administration Aircraft Data, and Apollo, Gemini, Skylab, and Space Shuttle data available to the general public. Contact EROS Data Center, U.S. Geological Survey, Sioux Falls, SD 57198.

Requests for information from these agencies should be as specific as possible. Requests for aerial photographs should state the purpose for which they are desired, and the area should be defined in a detailed description, sketch, or by the use of latitude and longitude coordinates.

Each state also has a State Geological Survey which provides relevant information. For the address of yours, ask a local science teacher or write to the Geologic Inquiries Group, 907 National Center, Reston, VA 22092.

•

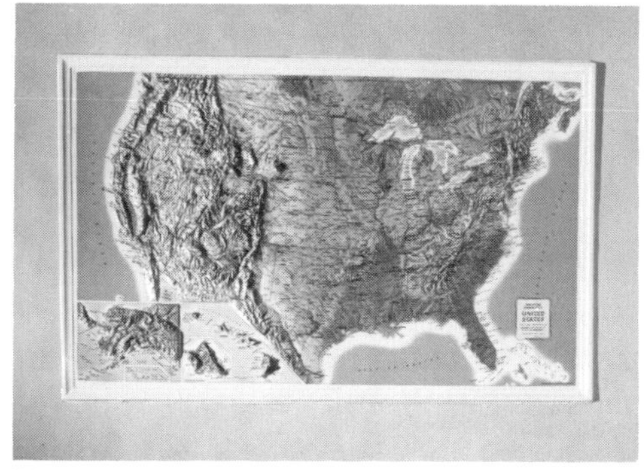

Fig. 50

Fig. 50. Accurate plastic relief maps from Geological Survey information are available through Hubbard Scientific. The most popular of these are the colorful 22″ × 35″ map of the United States as shown, and the 36″ × 20″ map of the world, both of which retail for $16.

For earth science fans planning auto or bicycle trips, there are a number of highway guides which highlight interesting geological features en route. These are designed for adults, but are worth sharing with the entire family.

A Geologic/Topographic Profile of the U.S. along Interstate 80 is available from Hubbard Scientific for $7. Book-length *Roadside Geology* guides to Arizona, Colorado, Texas, Oregon, Northern California, the Northern Rockies, and Washington, as well as guides to Waterton–Glacier National and Rocky Mountain National Parks, are sold through Mountain Press Publishers (P.O. Box 2399, Missoula, MT 59806) for $7 to $10 each.

If your intended destination is not listed above, write to the State Surveys in question. Many of them produce similar guides and information.

Mapmaking

Two companies produce items which give children hands-on experience with mapmaking.

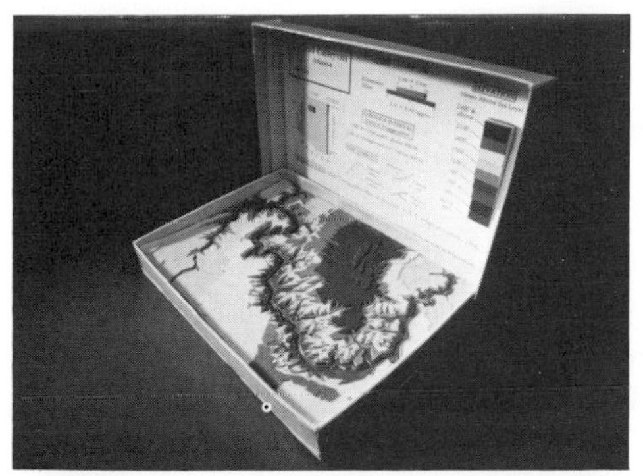

Fig. 51

Fig. 51. "3-D Grand Canyon" is just one volume in the build-your-own MAP-CRAFT LIBRARY OF GREAT PLACES. It comes with an overlay (not shown) which fits over the finished model, indicating features and sizes. Cutting out and gluing together layer upon layer of colored cardboard takes patience, dexterity, and a pair of surgical (or extremely sharp) scissors. Currently available are maps of Cape Cod, the Grand Canyon, the Great Lakes, Hawaiian Islands, Mount Rainier, and Yosemite Valley, which sell for approximately $9 each from Cartographics, 1849 S.W. 58th, Portland, OR 97221. Recommended for children age twelve and older.

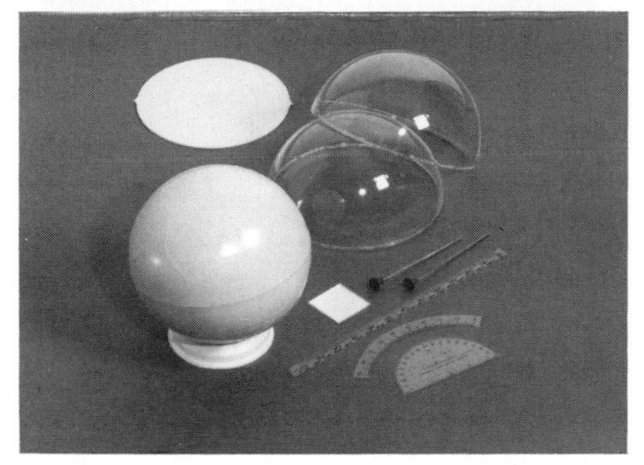

Fig. 52

Fig. 52. The GLOBE KIT is actually a blank model of the earth which enables older children to solve problems in earth measurement, astronomy, and seismology. Set includes an 8″-diameter earth globe, a transparent globe mold, internal earth cone disk, flexible kilometer scale, and an 8″ protractor, and retails for $13.25 from Hubbard.

Recognizing the need to produce interesting, accurate materials for middle-school students, the American Association for the Advancement of Science (1776 Massachusetts Ave., N.W., Washington, DC 20036) has reprinted a series of pamphlets entitled *Opportunities in Science.* The one on "Maps and Mapping," taken largely from *The Exploratorium Magazine,* a quarterly publication of the Exploratorium Museum in San Francisco, is excellent.

The U.S. Department of the Interior, U.S. Geological Survey, publishes a list of "Selected Books on Geology and Related Subjects" which marks titles as appropriate for elementary, junior-high, or high-school students.

General Selections

National Geographic Society, *Our Violent Earth* (National Geographic, 1982).

Outstanding photographs and straightforward text attune children of all ages to the traumas of our planet.

Raymo, Chet, *The Crust of the Earth: An Armchair Traveller's Guide to the New Geology* (Prentice-Hall, 1983).

Though intended for adults, this one works well for families. Lively and attractive.

For Young Children

Paola, Tomie de, *The Quicksand Book* (Holiday, 1977).

How quicksand is formed, why animals get stuck in it, and directions for making a version of the stuff in a tin can.

Rinkoff, Barbara, *Guess What Rocks Do* (Lothrop, Lee & Shepard, 1975).

A beautifully illustrated volume for young children which describes both ancient and modern uses of rocks and minerals. Also includes information on collecting.

Selsam, Millicent, *First Look at Rocks* (Walker, 1984).

An excellent beginning for the would-be collector.

Simon, Seymour, *Beneath Your Feet* (Walker, 1977).

Explains what soil is, how it's formed, and includes suggestions for related experiments.

For Grade-Schoolers

Asimov, Isaac, *How Did We Find Out About . . .* (Messner, 1978).

In this series Asimov describes how humans learned about events of scientific interest. Volumes on earthquakes, volcanoes, oil, and coal are particularly relevant to the fourth- or fifth-grade earth science buff.

Ruchlis, Hy, *How a Rock Came to Be in a Fence on a Road Near a Town* (Walker, 1974).

An appealing presentation for the seven- to ten-year-old.

For Middle-Schoolers

Bell, Neill, *The Book of Where, or How to Be Naturally Geographic* (Little, Brown, 1982).

An entertaining and instructive book on maps and mapping.

With a Little Help from a School

Teachers who write on school stationery to The Geologic Inquiries Group, 907 National Center, Reston, VA 22092, will be sent an extensive packet of teaching-related materials free. These packets are personalized by state—the package I received included geologic and topographic maps of New York, a list of water supply maps and reports, as well as more general information and bibliographies. The Geologic Inquiries Group responds to questions from the public (including teachers and students of all ages) about earth science and the research and publications of the Geological Survey. In short, children needn't wait until junior high to learn about the composition and characteristics of our planet.

Coburn, Doris, *A Spit Is a Piece of Land: Landforms in the U.S.A.* (Messner, 1978).

Clear explanations, good black-and-white photos, and lots of information are made more interesting by the inclusion of historical and etymological details.

A Poster

Using a time-line, artist John R. Stacy depicts Precambrian, Paleozoic, Mesozoic, and Cenozoic life and geologic activity in an informative and attractive 17″ × 22″ poster entitled "A Study in Time." Available from the American Geological Institute, 4220 King St., Alexandria, VA 22302, for $2.50.

Activity Guides

An EARTH SCIENCE ACTIVITY KIT from Educational Insights is available from Nasco for $7. It gives kids lots of suggested activities, and because these come on individual cards, they can easily be shared.

Tops, an outfit that produces ideas for teachers using low-cost or free materials, promises a new earth science series. Their catalog is free, their booklets sell for less than $10, and their suggestions for activities are usually excellent. Write Tops Learning Systems, 10978 S. Mulino Rd., Canby, OR 97013.

Earth Science Magazines

Although there are no earth science magazines specifically designed for children, *National Geographic World* and *Odyssey,* the space magazine, both include articles of interest to junior geologists.

Earth Science, 4/year, $8 (American Geological Institute, 4220 King St., Alexandria, VA 22302).

Intended primarily for the layperson interested in geology, this well-designed and informative magazine includes an excellent book review section featuring a wide variety of titles for young people. A calendar also lists meetings, shows, camps, and presentations of interest to non-professionals.

Lapidary Journal, 12/year, $12 (P.O. Box 80937, San Diego, CA 92138).

Designed for adult hobbyists, this periodical, available in many libraries, contains articles on gem history, collecting, and characteristics. More important perhaps are the many advertisements of specimens, rock and mineral tours, and a calendar of mineral society shows, activities, and meetings.

For Dinosaur Lovers

Dinosaur fever knows no age limits. These "monsters of ages past" (some of which were in fact quite small) have been immortalized in puzzles, stamp books, rub-ons, stamps, coloring books, templates, mobiles, T-shirts, models, posters, activity cards, and books. For an up-to-date list of dinosaur paraphernalia, send for your free "Dinosaur Catalogue": P.O. Box 546, Tallman, NY 10982.

For Younger Readers

Aliki, *Digging Up Dinosaurs* (Harper & Row, 1981).

Cartoons help explain the work of paleontologists in excavating dinosaur bones.

Carrick, Carol, *Patrick's Dinosaurs* (Houghton Mifflin, 1983).

Patrick's ability to relate dinosaur size to known objects and animals is a tribute to both the imagination and science.

Freedman, Russell, *Dinosaurs and Their Young* (Holiday House, 1983).
Unique in its approach for young children, this work offers a good summary of how dinosaurs gave birth to and cared for their offspring.

Lambert, David, *Dinosaurs* (Watts, 1982).
Fact-filled, colorfully illustrated, and easy-to-read descriptions of bird-hipped and lizard-hipped beasts.

Moseley, Keith, *Dinosaurs: A Lost World* (Putnam, 1984).
Dinosaur skeletons pop up next to representations of the animals; accompanied by a brief informative text.

Selsam, Millicent E., and Joyce Hunt, *A First Look at Dinosaurs* (Walker, 1982).
Readers enhance their powers of observation and learn the rudiments of scientific classification through this simple study of twenty-two dinosaurs.

Zallinger, Peter, *Dinosaurs* (Random House, 1977).
Excellent illustrations of twenty-five ancient flying and aquatic reptiles at a bargain price.

For Advanced Readers

Cobb, Vicki, *The Monsters Who Died: A Mystery about Dinosaurs* (Coward-McCann, 1983).
Fossil drawings serve as clues to dinosaurs' lives in this appealing, informative book.

Desmond, Adrian J., *The Hot-Blooded Dinosaurs: A Revolution in Paleontology* (Doubleday, 1976).
Evidence for the theory that dinosaurs were warm-blooded and quick-moving, and its implications.

Elting, Mary, *The Macmillan Book of Dinosaurs and Other Prehistoric Creatures* (Macmillan, 1984).
A lively account of ancient beasts and the paleontologists who study them.

Meyerowitz, Rick, and Henry Beard, *Dodosaurs: The Dinosaurs That Didn't Make It* (Harmony/Crown, 1983).
Kids with a well-developed sense of the absurd love this irreverent catalog of near-dinosaurs.

Simon, Seymour, *The Smallest Dinosaurs* (Crown, 1982).

Seven prehistoric species who lived in a world of giants for millions of years are described here.

Reference

Diagram Group, *Field Guide to Dinosaurs: The First Complete Guide to Every Dinosaur Now Known* (Avon, 1983).

Sattler, Helen Roney, *The Illustrated Dinosaur Dictionary* (Lothrop, Lee & Shepard, 1983).

Suitable for both reference and browsing, this handsome volume is wide in scope and comprehensive.

Books like these complement perfectly a trip to the Field Museum in Chicago, the Museum of Natural History in New York City or Dinosaur Monument Park in Vernal, Utah.

15 Chemistry, or, Lex Luthor Doesn't Live Here

In recalling the chemists to whom most young people have been exposed—*Sesame Street*'s Dr. Nobel Price, Dr. Bunsen T. Honeycutt of the *Muppets,* or Lex Luthor of *Superman* fame—it is no wonder that children often ask for chemistry sets, hoping to blow things up, create new life, and if all else fails, make an enormous mess. The producers of chemistry sets, on the other hand, actively work against these goals; they want their portable laboratories to be safe, intellectually substantive, and realistically challenging. For the would-be Dr. Doom, find a lab coat, a wig, and an out-of-the-way place to mix toothpaste and food coloring.

When he/she is ready, explain that chemistry is in many ways a difficult science to visualize; the homespun evidence that molecules even exist is indirect. Then point to those chemical reactions regularly evident at home: cooking eggs, or bleaching out stains. Help kids understand that these changes are due to changes at the molecular level. Whether or not you have back-pocket explanations of chemistry handy, there are certain assumptions made by chemists that can help children with this concept.

- Chemistry is not magic; reactions take place because something has changed.
- Changes, and the conditions that promote change, can be measured.

- Reactions are replicable and, in this sense, predictable.
- Because reactions are predictable, you can vary conditions, one at a time, and note the effects.

THE PLEASURES OF CHEMISTRY SETS

Although professional chemists and educators are interested in teaching chemistry as a process, they understand that the children for whom these sets are intended generally focus on product. Kids choose experiments because they want to make something happen: "I'll do the experiment that fizzes and turns blue."

To meet this expectation, American chemistry set manufacturers have turned to experiments they call "chemical magic," that is, the reactions challenge one's common sense expectations. In some sense there is real scientific value in these "gee-whiz" experiments. They prove to children that things are not always as they seem, and that scientists have more than a common-sense understanding of the physical world. The designation "chemical magic" is, however, bothersome. It implies that the scientist has fooled his/her audience, and that the explanation of these startling events is simple, and purposefully obscured. Experiments of this sort also liken chemists to cooks using prepackaged products—one needn't have any understanding of the function of baking powder or eggs to make a "modern, just add water" cake.

Sets designed outside the U.S., by contrast, seek a more classical understanding of chemistry. The kits are designed to make both ordinary and extraordinary reactions interesting and meaningful. Rusting, for instance, is seen in these booklets as a fascinating chemical reaction. "Trendy chemistry"—hydroponics, ecology, and chemical magic—is generally absent. These sets are for disciplined and interested children, not for someone likely to thumb through a guide asking, "Do I want to make metal rust?" "Do I want to make milk curdle?"

At their best, chemistry sets teach several important things. First, the experiments (which are in fact demonstrations) provide information and develop concepts that may be generalized and applied. A child who has burned copper salts and noted their green color may realize that an identical process is used in fireworks. Similarly, experiments with nitrogen-deprived plants may help an observant kid account for the yellow color of sickly crops.

In addition, these sets allow children to understand that there is a physical aspect to doing science. One learns to measure carefully, to read numbers accurately, and to burn substances safely. This "chemical cooking" is attractive, even exciting, to certain young people. The same interest in handling materials is apparent in working scientists—a biochemist who works with recombinant DNA claims that "having good hands" is an important part of being successful in his field.

Chemistry sets also allow product-oriented children the satisfaction of making what they started out to make. And if the desired reaction didn't occur, the satisfaction of going back, and this time getting it right.

Finally, chemistry sets invite independent and self-motivated kids to pursue knowledge as they wish—working furiously for several days or weeks, then ignoring the subject for perhaps months on end. Young people choose experiments for different reasons—sometimes they want a quick and easy demonstration that they know will work, while at other times they seek more complicated ventures. A well-designed instruction booklet invites a diversity of exploratory moves.

HANDS-ON

Pre-Chemistry Set Ideas

There are many reasons why one might not want to begin children interested in chemistry with a chemistry set. First, there are no such items designed for children under age nine. Manufacturers are appropriately afraid that warnings of danger may go unheeded, and that explanations are far too abstact for small children to follow. Some parents may not be financially ready to invest in the set of their dreams, while others may be put off by the whole concept of pre-fab science.

For any child genuinely interested in chemistry, the following measuring tools are good choices: thermometers, graduated plastic beakers, measuring spoons, and a balance. With them kids get used to the idea that chemistry is about varying amounts and conditions, become more dexterous in handling lab tools, and feel, appropriately so, that they have "real" equipment. (There are several books, listed below, that give ideas on what to measure and for what results.)

Corked glass bottles for saving solutions, and litmus paper for testing for acids and bases, are also appropriate and useful gifts for children of all ages.

Safety habits can also be established by explaining to children why you get them disposable cups, stirrers, and so on—equipment used for chemistry should not be used for any other purposes, especially eating. Safety goggles are also fun and important for children.

Chemistry Sets: An Evaluative Guide

Six companies distribute chemistry sets in the United States today. In alphabetical order these are:

Battat and Suitcase Science import from Japan the same inexpensive ($5 to $6) kit featuring five unique experiments and the equipment to perform them. Although no explanation of the "chemical magic" is offered, the experiments are fun, they work, and they are not replicated in any of the other sets.

E.D.I., an American firm, sells a chemistry set called POWERTECH, which is produced in Israel and costs $30. Although the manual includes many typographical errors, and the packaging of the chemicals is not as flashy as in other sets, this kit wins high praise. The organization of the manual is thoughtful, and its author, Peretz Mahler, writes with enthusiasm and humor. Experiments are arranged sequentially: a concept is introduced, the principles discussed are demonstrated by the child, and a gee-whiz experiment takes the user on to further explorations. Acid/base chemistry and chromotography are both clearly explained. More to the point, both child testers and professionals chemists were pleased with the experiments and explanations provided.

Consumers Beware

In trying to decide which chemistry set to buy, consumers generally look at the number of experiments offered, the number of chemicals included, the quality and quantity of glassware and equipment, and the overall packaging. This approach can be deceptive—here's why.

- Manufacturers, aware of the urge to count experiments, often offer "fillers," experiments that use nothing from the set. For example, a set may include adding vinegar to milk as an experiment, although one needn't buy the set in order to curdle the diary product. Other sets simply reverse experiments in order to inflate numbers, for example: "Experiment 1: Add vinegar to milk. Experiment 2: Add milk to vinegar. Experiment 3: Add vinegar and milk at the same time."
- The number of chemicals is also not a sure-fire way of judging the set's worth. Sets can include, for instance, 5% acetic acid, which is nothing but household vinegar, or sodium bicarbonate, which is another name for baking soda. A set which includes a few unique and multi-purpose chemicals is better than one that has lots of substances available in any kitchen. *Note:* To cut costs, manufacturers often mix expensive chemicals with less expensive ones, for instance cobalt chloride with table salt. These mixtures work just as well as the purer forms in the suggested experiments.
- Glassware and equipment may also be added as an advertising come-on. Disposable cups can work as well or better than glass beakers; a balance used in only one experiment might be better left out completely.
- There is little correlation between the care manufacturers take in packaging items and the quality of the items themselves. Pretty packaging does not necessarily mean good directions and satisfying experiments.

In short, whenever possible, read through the instruction booklet before buying.

Ideal, a toy company now owned by CBS, produces three chemistry sets: CHEMCRAFT **50** PROJECT MAGIC LAB for $9, CHEMCRAFT **500** PROJECT SCIENCE LAB for $15, and CHEMCRAFT **1000** PROJECT SCIENCE LAB for $24. The larger sets include all the experiments in the smaller kits. Although more traditional topics such as acid/base chemistry and simple displacement reactions are included in the larger sets, all the CHEMCRAFT manuals are liberally sprinkled with experiments reviewers term "chemical magic." Food chemistry and ecology experiments are also included. With few exceptions, experiments are not arranged sequentially, and children generally wander freely through the manual, deciding what they want to happen next. Chemical explanations are brief, and in the chemical magic experiments, nonexistent.

The instruction manuals in the CHEMCRAFT sets have recently been revised. Make sure that you get the most recent version; earlier editions include hard-to-follow directions, important typographical errors, and a poorly organized and unclear presentation of fact. Childproof caps, an excellent idea in theory, are extremely difficult to open; in two instances the cap tabs completely ripped. (If this happens, have an adult pry open the bottle and store the contents in a more accessible jar.) Except for the balance, the equipment is very good. The balance, a poorly designed and hard-to-use instrument, was required for only one experiment in the 500 kit and the 1000, although it can be used instead of the measuring spoon. The set is nicely packaged; it comes with a set of handy storage racks for chemicals and equipment, and includes an order form for replacing chemicals and equipment. For children age ten and up.

Merit, a British manufacturer, has produced six chemistry sets which range in price from $27.50 to $150. Distributed in this country by Polk Bros., a firm owned by two brothers whose philosophy is not unlike that of the original A. S. Gilbert, these sets can be purchased directly from Polk Bros. Hobbies International, 314 Fifth Ave., New York, NY 10001, and through selected museum stores. The sets include little "chemical magic" and food chemistry; rather they help children explore classical experiments on water, gases, chromotography, etc. Explanations provided at the back of the manual are clear. Britishisms used in the manual may be unfamiliar to an American audience—"spirits," for instance, is another name for alcohol. The beginning set could be used by a nine-year-old, while the advanced set contains glassware and experiments not unlike those found in a high-school lab.

Natural Science Industries, a family-owned firm which originally specialized in butterfly collection and mounting equipment, imports a line of chemistry sets from Scotland known as the **THOMAS SALTER LABS.** Four sets are currently available through toy and department stores: a junior chemistry set retails for $9.50, a basic set of 82 experiments for $26, an intermediate set of 96 experiments for $32, and an advanced set of 133 experiments for $50. The number of advertised experiments is accurate, the chemistry is not at all trendy, and the explanations are thorough and clear. The equipment in the kits is excellent, in many instances of professional quality. Experiments are classic and safety precautions clear—users study galvanization, rusting, and experiments to characterize compounds and gases (such as generating a relatively harmless amount of chlorine gas from solid calcium oxychloride).

The N.S.I./Salter set has two advantages over similar lines: for an imported set it is not terribly expensive, and it includes a recorded introduction to chemistry by an enthusiastic British popularizer of science, Johnny Ball. Ball is great. He leads the child through what might, under other circumstances, be a boring experiment and makes the ordinary fascinating. In so doing he helps children understand that chemistry is more than a series of magic tricks. For those who are reluctant to begin, this set could work wonders. N.S.I./Salter also includes "translations" of Britishisms for American audiences. Recommended for age ten and up.

Skilcraft, an American science and hobby company, produces five chemistry sets called the CHEMLAB line: a **BEGINNER'S CHEMLAB** (age nine and up) which retails for $9; an **INTERMEDIATE CHEMLAB** (nine and up) for $15; an **ADVANCED CHEMLAB** (ten and up) for $19; a **SENIOR CHEMLAB** (ten and up) for $27; and a **GRADUATE CHEMLAB** with carrying case (actually the Advanced set better packed) for $35.

The Advanced set was reviewed. It advertises and includes "chemical

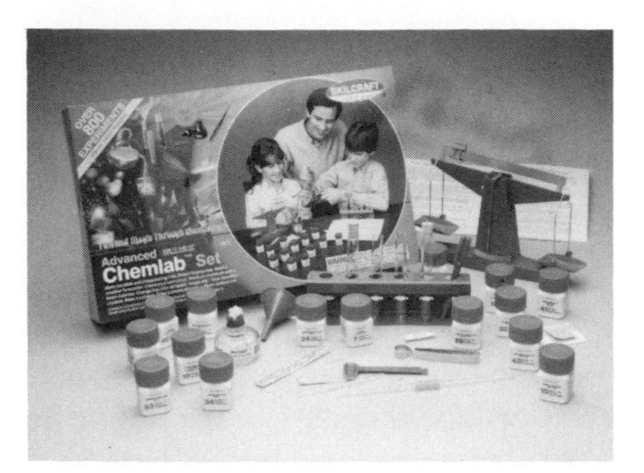

Fig. 53

magic," a section on hydroponics, and other trendy and attractive topics. In addition, experiments examining physical changes, the chemistry of specific elements (such as oxygen and sulfur), dyes, acid/base chemistry, electrochemistry, reactions that form precipitates, chemistry of solutions, and other classic experiments are included. The Skilcraft explanations, especially in the newly revised manual, are detailed and exceptionally well organized. Although users can skip from section to section, within a particular unit they should work sequentially. Skilcraft does not fill orders for additional chemicals or equipment, but does refer users to local hobby shops for replacement parts. Recommended for age ten and up.

Tips

- If you are unsure about whether a chemistry set would work well for your child, begin with either the Battat/Suitcase Science set, the N.S.I. junior chemistry kit, or the Skilcraft beginner set, none of which requires a large outlay of cash.
- Sets from the same countries tend to have much in common.The Skilcraft and Ideal sets include many of the same experiments. Similarly, the Salter and Merit kits have some overlap. But with little replication of experiments one can move, for instance, from the Battat set to the POWERTECH set, to the advanced CHEMLAB, to the advanced THOMAS SALTER LAB.
- For a child fascinated by gee-whiz science, I would buy the Skilcraft kit and try to move him/her from gee-whiz to more systematic experiments. For a child with a more sustained and lively interest, I recommend the POWERTECH or SALTER kit.
- For a nine-year-old with serious scientific interests, begin with the POWERTECH set. For a fourteen-year-old with a good science background, start with an advanced SALTER, CHEMLAB, or MERIT set.
- The best buy in a science kit is Skilcraft.
- Try the kitchen chemistry books (listed below) before purchasing an expensive lab.
- For those who fondly remember their old GILBERT CHEMISTRY SETS—

Fig. 54

rows of chemicals in a gleaming metal box—be warned: there is nothing comparable on the market today.

Fig. 53. The prize for the best organizational scheme goes to Skilcraft's CHEMLAB. These sets successfully cover the widest range of topics.

Fig. 54. First place for equipment and the teaching of technique goes to N.S.I./Salter.

Fig. 55. The POWERTECH manual wins the best-instructor award. Users describe the experiments as "rewarding."

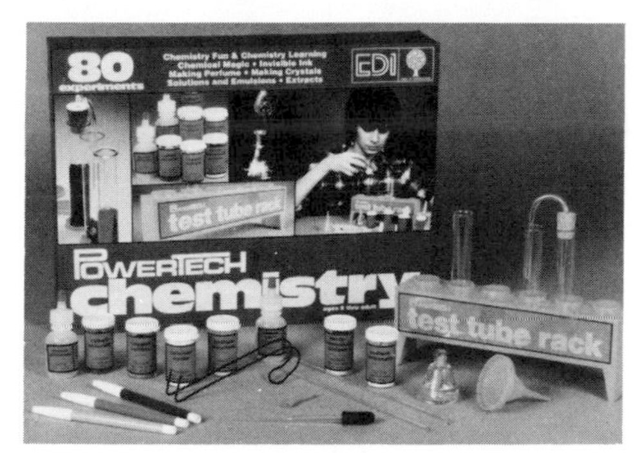

Fig. 55

Used Chemistry Sets

Another inexpensive alternative for the penny-wise parent is a used chemistry set, either a hand-me-down or a porch sale bargain. Older chemistry sets, especially those from the 1950s and 1960s, are better than many sets available today, although users had the potential to make a number of dangerous compounds. If you find a used set, consider the following: Chemicals can be replaced quite easily—the Perfect Parts Company line is available in most hobby stores, and catalog houses like Merrill Scientific also sell their own extensive line—but you should be sure that the apparatus (burners, scales, glassware) are in working order. Most important, *never* buy a used chemistry set without all the manuals that were included in the original. Also, because some older sets were produced at a time when less was known about links between chemicals and disease, make sure that *no* chemicals listed on page 201 are used by your child.

Chemistry Miscellany

Electroplating Kit

A North Carolina-based wholesaler, Science and Nature Distributors, makes an electroplating kit which retails in museum shops and science stores for about $3.50. This kit is, in fact, a single experiment which allows kids to put a metal coating on the surface of objects. It's simple, it works, and it's fun.

Periodic Table Puzzle

Hubbard Scientific puts out a puzzle of bogus elements that students arrange in much the same way Mendeleyev organized the periodic table we now use.

Fig. 56

The puzzle isn't "fun" in the sense Monopoly is fun, but it is a genuinely challenging and interesting experience for an older child who has completed a number of chemistry sets. An exercise in logic, as well as chemistry. *Note:* After working through this puzzle with your children, give it to a chemistry teacher for use in school; it isn't designed for extended play.

Fig. 56. A novelty item as well as a chemistry experiment, the TWO POTATO CLOCK runs on electricity generated by a chemical reaction—the acid in the food reacts with the metal probes to generate a flow of electrons. Lemons, apples, cola, house plants, even beer, can be used as an energy source; users experiment to see which substances make the digital readout work most reliably, and for the longest period of time. The TWO-POTATO CLOCK from Skilcraft sells for approximately $15 in most toy and museum stores.

A simple distillation device allows users to make their own perfume. It's a five-step process requiring readily available substances (lard, denatured alcohol) and an adult's help; $22 from Edmund Scientific.

Crystal Growing Kits

Crystals are regularly repeating patterns, formed by the solidification of elements or chemical compounds. Because the shape of the crystal reflects what is happening with the substance on a molecular level, and because there are a limited number of crystal patterns that form, crystal structures provide important information for chemists, geologists, and physicists. In addition, many crystals are colorful and complicated. In each of these kits, the user dissolves a compound in water, and watches the crystals form and grow larger. When the reaction is completed, the child can display and study the resultant forms.

Science and Nature Distributors produce a simple crystal growing kit which retails for $4 at museum shops.

The most complete crystal growing line is created by a German company, Kristallzuchtung, which sells here under the name SPACE AGE CRYSTALS. Recommended for age twelve and older (with a parent's ongoing supervision, a younger child would enjoy these products), single- and multiple-substance

WARNING . . . WARNING . . . WARNING

Many of the chemistry experiment books still available in local libraries encourage children to work with chemicals that are, by 1986 standards, extremely **dangerous.** For instance the first edition of Virginia Mullin's *Chemistry Experiments for Children* called for benzene, a known carcinogen. Although revised versions have dropped this experiment, older editions are still in circulation. More recent publications have also not been as careful as they should be in checking toxicity levels of substances. For these reasons chemists recommend that **EXPERIMENTS USING ANY OF THE FOLLOWING CHEMICALS SHOULD DEFINITELY *NOT* BE UNDERTAKEN AT HOME.**

Benzene, a.k.a. benzol

Carbon tetrachloride, a.k.a. carbon tet, a.k.a. cleaning fluid

Formaldehyde, a.k.a. formalin (preserved specimens are pickled with very dilute formaldehyde. Even so, keep these solutions well stoppered and avoid breathing fumes.)

Lead nitrate

Mercury, a.k.a. quicksilver

Paraformaldehyde

For a detailed description of possible effects or to check other "suspect substances," see the latest edition of *The Merck Index* (Merck & Co., Inc.) available in most public libraries.

kits are available. Each creates crystals of a different pattern and color. Single-substance kits cost $8, the four-substance set retails for about $22, and the large kit, which contains twelve substances and creates up to 10″ specimens, sells for $54. Sets are available directly from Space Age Crystals, P.O. Box 44287, Tucson, AZ 85733, through science catalog houses, or in better toy stores.

Scott Resources (P.O. Box 2121, Fort Collins, CO 80522), known primarily as a mineral supply company, also sells crystal growing kits. Their line is slightly more expensive ($14 to $67) because they cater to the school market and provide chemicals in quantity.

IN PRINT

For Young Children

Gans, Roma, *Millions and Millions of Crystals* (Crowell, 1973).
Crystals are large representations of molecular activity. Gans tells you about this and more.

Goldin, Augusta, *The Shape of Water* (Doubleday, 1979).
Gases, liquids, and solids are presented in this straightforward and accessible guide.

Haines, Gail K., *What Makes a Lemon Sour* (Morrow, 1977).
Explains and provides experiments related to bases and acids. The perfect accompaniment to a gift of litmus paper.

For Grade-Schoolers

Gardner, Robert, *Kitchen Chemistry* (Simon & Schuster, 1982).
Not a book about food chemistry, but actually a series of experiments, using paper, candles, and air, performed in the kitchen. The concepts it so ably teaches—solubility, crystal growing, density, diffusion—are known as physical chemistry.

Shalit, Nathan, *Cup and Saucer Chemistry* (Grosset & Dunlap, 1972).
Don't be put off by the publication date on this one; the experiments are clearly presented and satisfying.

Simon, Seymour, *Chemistry in the Kitchen* (Viking, 1971).
Easy-to-find materials are used to answer common questions employing hands-on techniques. Have you ever wondered what happens when you beat eggs?

For Middle-Schoolers

Asimov, Isaac, *How Did We Find Out about Atoms?* (Walker, 1976).
The clarity of the explanations and the historical perspective offered in this volume make it a good choice for the advanced reader.

Cherrier, François, *Fascinating Experiments in Chemistry* (Sterling, 1978).

These high-interest experiments—making laboratory glassware, invisible inks, small fireworks, and more—require *careful supervision.*

Chisolm, Jane, and Mary Johnson, *Introduction to Chemistry* (EDC Publishing, 1983).

Don't be deceived by the cartoon-like graphics—this is a sophisticated introduction to the subject which provides theory as well as experiments.

Cobb, Vicki, *Chemically Active: Experiments You Can Do at Home* (Lippincott, 1985).

Readily available materials are used to illustrate important and interesting chemical principles; an excellent chemistry set in a book. *Note:* Cobb uses the word inflammable in reference to things that burn. Though technically correct, warn children that inflammable is synonymous with flammable.

———, *Science Experiments You Can Eat* and *More Science Experiments You Can Eat* (Lippincott, 1972 and 1979).

These food experiments are all well done—informative, interesting, and edible.

Davis, Edward E., *Into the Dark* (Atheneum, 1979).

The best young adult step-by-step guide to darkroom technique and equipment to date.

Keller, Mollie, *Marie Curie* (Watts, 1982).

Although the interpretation of Curie's American trip is debatable, Keller's biography is in other respects well formulated and fascinating.

Mullin, Virginia, *Chemistry Experiments for Children* (Dover, 1968).

All the information you need to build your own, relatively elaborate, chemistry set. Of the fifty recommended chemicals, thirty-eight are easy to find in drugstores, hardware shops, and supermarkets. *Note:* The trick to hunting down chemicals is knowing both their common and scientific names. For instance, potassium hydrogen tartrate is the technical name for cream of tartar. Check an unabridged dictionary for this kind of information. Be sure that all glassware used is labeled "heatproof" or Pyrex.

Zubrowski, Bernie, *Messing Around with Baking Chemistry: A Children's Museum Activity Book* (Little, Brown, 1981).
By baking a series of little cakes, the author clearly demonstrates the scientific method, while never "talking down" to his readers.

A Magazine

Chemmatters, 4/year, $6 (David A. Daniel, American Chemical Society, 1155 16th St., N.W., Washington DC 20036).
This periodical designed for high-school chemistry classes may well be accessible to a junior-high-schooler with interests in chemistry. Articles are relevant (suntanning and polarized light, liquid crystal displays, etc.) and informative.

16 Phascinating Physics

In high school we learned that physics is the science that comes after biology and chemistry, a subject accessible only to the most "fit" scientific minds. This belief is somewhat ironic, given that toddlers regularly experiment with, and naively define, what objects in this world can and cannot do. Moreover, a knowledge of the laws of matter and energy is basic to an understanding of all science and technology. This is not to say, of course, that young children can make sense of the mathematical formulae upon which contemporary physics is based. Rather, early physics should be designed to give kids experience with, an intuitive sense of, and an experimental strategy that serves them well when they do encounter the mathematical descriptions used by scientists and science teachers.

Appearances suggest that there is a dearth of physics toys on the market. Remember, however, that physics is essentially concerned with the physical properties and composition of everyday things; in fact, your home, the hardware store, the toy store, and the great outdoors are replete with items to teach about mechanics, acoustics, optics, heat, and magnetism. Because physics represents a way of looking at phenomena, the books in this section are particularly useful for the child who wants to know more about physical laws.

HANDS-ON

Physics Sets

Fig. 57

POWERTECH PHYSICS from E.D.I., the only physics set currently sold in the U.S., is available through science stores, toy shops, and science catalogs for approximately $40. A nice variety of experiments on gases, liquids, static electricity, motion, light, heat, and sound are included. Explanations are good, and the way in which mathematical formulae are introduced is impressive. The equipment provided is somewhat troublesome—balloons that don't blow up, items with a dual function that often work in only one situation (for example, a wheel designed to double as a pulley and an air puck that was too heavy to ride the stream of air), although these devices are easily replaced (our dime-store balloon worked). More troublesome is the absence of certain parts listed in the manual. Still, the manual is well conceived and written, and with the addition of easy-to-find materials, this set provides a nice introduction to the field. Recommended for age eight to adult.

Fig. 57. Edmund's SCIENTIFICS, a line available through Edmund and many museum shops, is in fact a series of small physics sets. Moderately priced, these hands-on items are packaged with an explanation of the scientific principles involved and sometimes a brief biography of the physicist who first explained the ideas demonstrated. Included in the SCIENTIFICS line are weather, magnetism, color, mirror, and illusion kits; purchased direct from Edmund, you get all five sets for the bargain price of $10.

Fig. 58. A JUNIOR DYNAMICS KIT from Natural Science Industries ($7.50) provides a good hands-on introduction to the physical sciences. Johnny Ball, a lively British popularizer of science, adds markedly to the fun with his tape-recorded lesson.

Physics Tools

Edmund sells an impressive collection of lenses, optical equipment (such as optical benches and lens brushes), fiber optics, polarizing sheets, color filters, and lasers. Most of these materials are not for the novice, but could be useful in science fair projects or advanced experiments.

Fig. 58

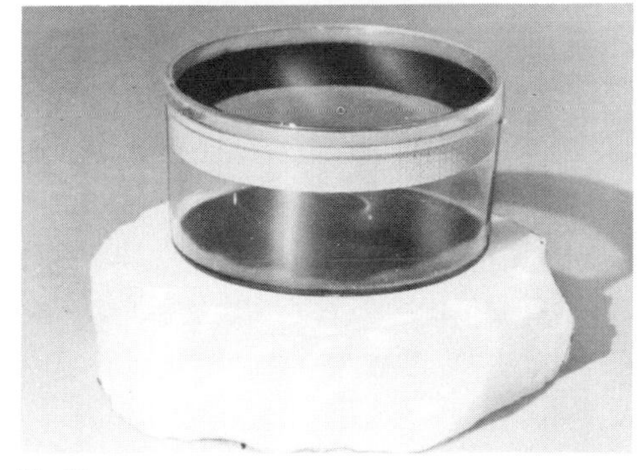
Fig. 59

A. Jaegers (6915 Merrick Rd., Lynbrook, NY 11563) also sells a wide selection of lenses and mirrors, many of research quality. For a catalog send 25 cents to cover handling.

In addition, surplus dealers (such as Jerryco) regularly carry lenses, prisms, small scopes, and other goodies at bargain prices; what is available, however, changes from month to month.

A hand-held spectroscope, a neat device for examining spectra, is available from Hubbard Scientific for $2.

The JUPITERSCOPE, an inexpensive round, flexible prism made with a laser, is available in most science museum shops. Even young children can enjoy seeing common sources of light broken into spectra, while older kids use the lens as a camera attachment. An excellent stocking stuffer.

Fig. 59. Radioactive materials are generally not for playful experimentation. There is one interesting exception: the CLOUD CHAMBER KIT from Hubbard ($4.25). Using a plastic cloud chamber, a piece of uranium ore, alcohol, and dry ice (from a nearby ice cream store), one can observe the tracks of alpha and beta particles. The uranium ore is kept in a container; it is essentially safe as long as you follow package instructions carefully.

Atom smashers and other important physics equipment are on display in several of the National Laboratories. Trained tour guides do an excellent job of explaining the instruments' purpose and design.

Magnets and Magnetism

Magnetism is a perfect example of both the accessibility and the genuine complexity which characterize physics. Very young children delight in playing with magnets—for some mysterious reason these objects pull some metals toward them, have no apparent affect on non-metals, and repel each other. Because magnets defy commonsense expectations they are perfect instruments for scientific investigation—kids can describe what magnets can and can't do. On the other hand, magnetism is an extremely difficult concept to understand and explain. With young children, wait for questions to arise, and if and when they do, find a good book to help you answer queries. In short, don't think of magnets as items for tykes; magnetism could well serve as the basis for an advanced science fair project.

A **MAGNET KIT,** nicely packaged and organized, is distributed by Battat and Suitcase Science, and sold in toy stores and museum shops for about $6. Recommended for children ages four to eight.

The magnet magnate has to be Edmund Scientific, which sells thousands of different-shaped magnets at reasonable prices. For further information on these, see the Edmund Catalog for Industry and Education (see page 119).

Edmund also sells a magnet kit in its **SCIENTIFICS** line. It includes several ideas for experiments and is reasonably priced at $4. **NATURE'S MAGNET,** the lodestone, can be found in Edmund's **NATURIFICS** line ($3). These kits are all available direct from Edmund or from science and museum shops.

A more elaborate magnet set is sold by Lakeshore. Intended for school use, this $35 kit contains a variety of magnets, a quantity of iron fillings, a compass, and an excellent selection of ideas for experiments.

Fig. 60. **MAGNASTIKS,** the most artistic of the magnet kits, includes ninety metal magnets that kids use to build magnetic sculptures and designs. Complete with a handy storage case, this set, recommended for children age five and older, sells for $18.

Fig. 60

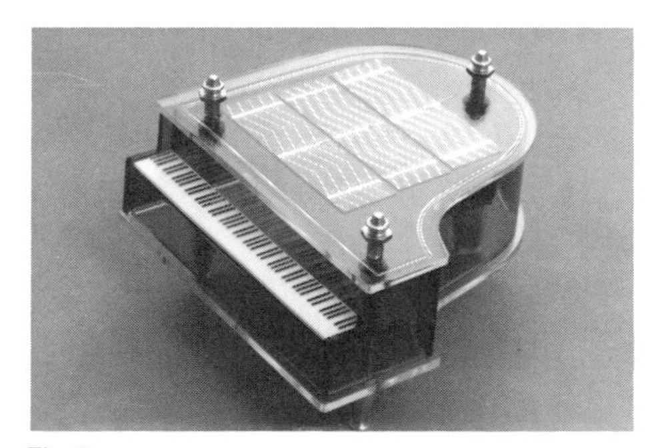
Fig. 61

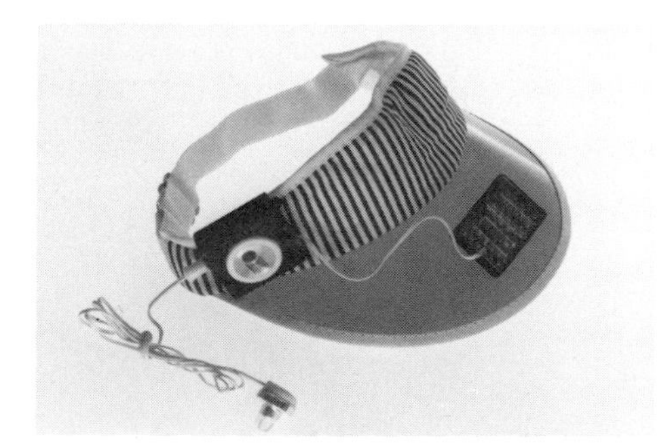
Fig. 62

Solar Energy Kits

The most artistic, safest, and easiest to use of the solar energy items is light-sensitive paper, often sold under the name SOLARGRAPHICS ($5). Users arrange opaque items on these blue sheets, enjoy the sunshine for about 5 minutes, run the paper under clear water to "fix" the image, and surprise!, where the light didn't hit, an image appears. Learning Things sells this light-sensitive paper in bulk.

A series of solar-powered model books published by Price / Stern / Sloan (Los Angeles) that use a solar cell to move punch-out figures and machines are available, complete with solar cell, from bookstores for $10.

The principle that the light from the sun can be used for energy is demonstrated in a variety of fascinating novelty items—(fig. 61) a solar-drive music box ($20), (fig. 62) a hat/radio run by a solar cell ($30), or (fig. 63) a glass radiometer which spins faster as brightness intensifies ($8). These gadgets can be ordered from a science catalog (the Energy Sciences catalog carries an especially good line) or purchased directly from a science store or museum shop.

Fig. 63

An Edmund SCIENTIFICS PLUS KIT which contains a solar cell connected to a small dc motor used to spin a propeller or wheel is available for $11.

A larger SOLAR ENERGY KIT from Energy Sciences includes a 1¾″ × 2⅞″ panel, 1.5 volt dc motor, and a book explaining the principles upon which solar energy is based. Available through Energy Sciences or in novelty and hobby shops for $16.

Both Edmund Scientific and Energy Sciences sell a large selection of silicon solar panels which can be used in a variety of photovoltaic experiments. For project ideas, see *The Energy Sciences Project Book: 33 Solar Power Projects for Hobbyists,* available from Energy Sciences.

The SOLAR COLLECTOR, a single experiment from Battat, provides a simple demonstration of the sun's power. The set is modestly priced ($5 to $6), well designed, and recommended for children ages six to ten.

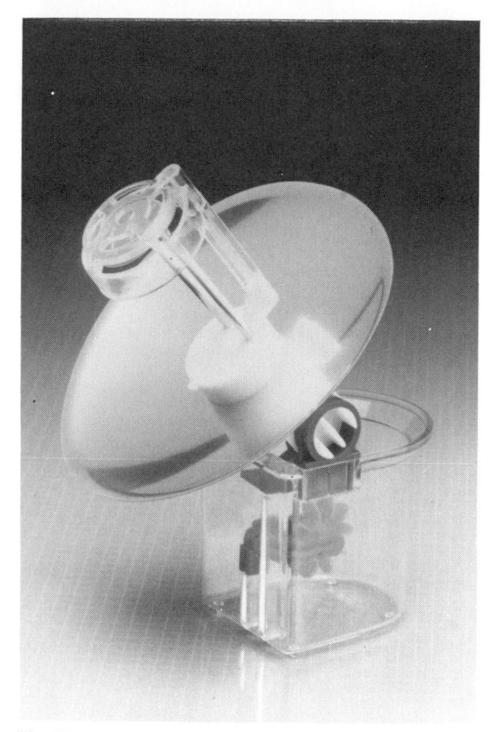

Fig. 64

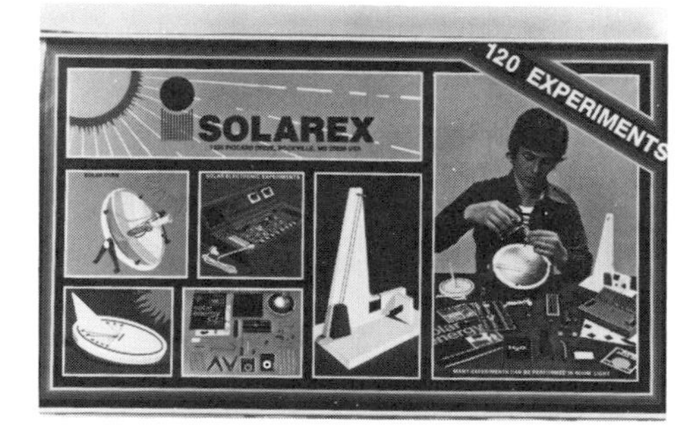

Fig. 65

Fig. 64. The Suitcase Science **SOLAR COLLECTOR,** from Small World Toys, recommended for children ages six to twelve, is also inexpensive and fun.

My favorite solar kit, **SOLAR WORKS,** may still be found on the shelves of toy stores, although it is no longer being produced by Ideal. This easy-to-build collector comes with an attractive, well-written manual, and users describe the experiments as satisfying and fun—cooking a hot dog, making solar potato chips, distilling pure water from salt water, and the ever-popular burning a hole in paper. Highly recommended for children age eight and older.

The E.D.I. **MINI-LAB SOLAR ENERGY KIT,** which retails for $7, uses the sun's energy to explore basic science (whereas a kit like **SOLAR WORKS** explores solar technology). Experiments include calibrating a thermometer, brewing tea, locating the focal point of the solar collector, and building additional solar collectors. While the experiments are excellent in theory, directions are sloppy (experiment 21 is listed before experiment 20, a change in order which *does* make a difference). With adult help and ingenuity this kit is recommended for children age eight and older.

Fig. 65. The *crème de la crème* of solar kits is sold for $30 under two labels, Energy Sciences and Tree of Knowledge/E.D.I. The manual is excellent, the explanations of demonstrations detailed, and numerous practical applications of solar technology are included. Experiments explore the science and technology of solar heating and include evaporation demonstrations, sundials, solar cells, and systems for calculating the energy consumed by a hot bath, and charting degree days. The photovoltaic experiments, a series of simple circuits powered by a solar cell, are also rewarding. (1) Unlike most of the electronics sets, the capacitors, transistors, and resistors in this kit must be wired into a plastic board—a difficult task for awkward hands; (2) an able thirteen-year-old and two Ph.D. scientists could not get the last experiment, a perpetual-motion pendulum, to work. In general, however, this is a sophisticated and ambitious set for children age twelve and older.

Fig. 66. The **SCIENCE FAIR** line, available at Radio Shack stores, features a **SOLAR POWER LAB** ($17), actually an electronics set, with four solar cells to generate electricity. Directions teach you how to build a radio, logics circuits,

Fig. 66

and a Morse code key. The procedure for connecting wires—using spring contacts in a fixed board—is easy to use. A nice introduction to electronics for children age nine and older.

Sun-powered **STIRLING ENGINES,** available from Solar Engines (4200 E. McDowell, Phoenix, AZ 85008), unlike conventional internal combustion engines, require external heat to run. The heat drives a piston, the piston drives a wheel, and the engine drives anything you can design for it. These energy-efficient and quiet engines may well be the power-plant design of the future. *Warning:* The engine gets hot and stays hot as it runs, so parental assistance is definitely recommended. These beautifully machined items, which are sold assembled or unassembled, cost $27 to $64 and are recommended for age ten and up.

A **PORTABLE SOLAR COOKER,** which really works, is available through the Edmund catalog for $23.

What You Need to Know About Solar Experiments

- Solar energy kits which use a parabolic mirror to collect and focus the sun's energy pose a danger for careless children. Concentrated sunlight can start fires and can hurt the user's eyes if directions are not followed precisely.
- The success of solar experiments largely depends upon the weather, time of day, and the location of the collector—noon on the Mojave Desert is ideal. Young experimenters in Ohio were disappointed, for instance, that they could not boil water with the solar collector, although the temperature of the water certainly rose. To maximize solar capacity, you may have to move the collector to keep it in sync with the sun.
- Don't be too impressed by the number of experiments the kits claim to contain. For instance, one set counts the suggestion that users see if a light bulb throws off heat as a separate experiment.

Toys for Experimenting with Physical Principles

As mentioned earlier, physics is evident in the working of many toys and everyday objects—balloons, tops, drinking straws, pinwheels, eggbeaters, and bicycle pumps. Here are a few excellent items you might have missed.

For All Ages

Fig. 67. The Tedco/Chandler gyroscope, a metal wheel which rotates freely about an axis, is a fascinating hand-made physics toy. Available in most museum shops, at local fairs, and in science stores for about $5, this instrument gives meaning to words like angular momentum, precession, and torque, while providing hours of amusement.

For Young Children

Fig. 68. KNOCKY, distributed by Creative Playthings (a slightly more expensive, wooden version of this same toy called POUND A BALL is available through other manufacturers), helps children understand that the energy and order with which one strikes the balls affects the outcome. This cause-and-effect toy is recommended for children ages three to five; $12 from Childcraft.

Fig. 69. BUILDA HELTA SKELTA from International Playthings ($17 in better toy stores and catalogs) includes interchangeable chutes which children arrange as they see fit. As the marble rolls through the maze they've built, users learn logic, manual dexterity, and problem-solving skills. *Not* for children who put marbles in their mouths!

Fig. 70. A $4 pump from International Playthings, unlike most bath toys whose opaque gadgets mysteriously spit and spin, shows the entire workings

Fig. 67

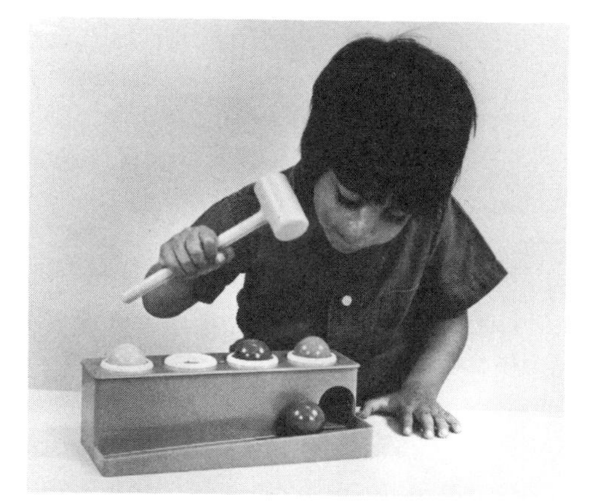

Fig. 68

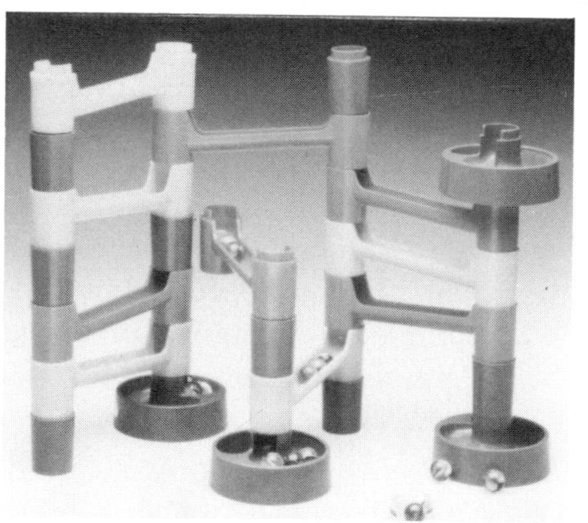

Fig. 69

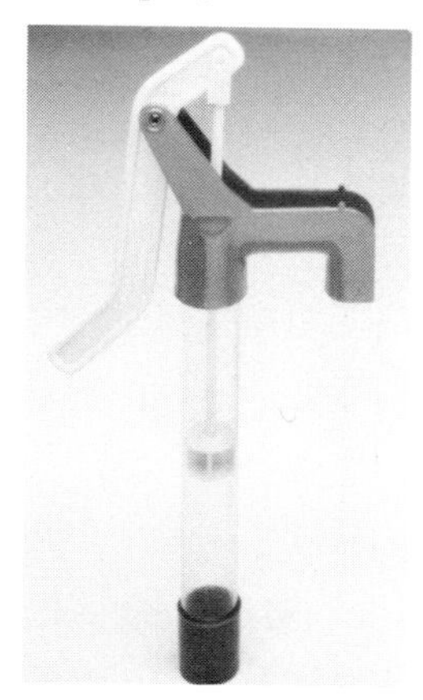

Fig. 70

of the simple machine. Moreover, the three-piece system can easily be taken apart; if put back incorrectly, the pump simply does not work.

Kaleidoscopes

Kaleidoscopes, a toy that combines physics and art, can be purchased from most toy stores. Although several do-it-yourself kaleidoscope kits which supply the requisite mirrors and tube have been available, none is at present. Be on the look-out for these, however. Making your own instrument is not at all difficult, teaches important concepts, and breeds that feeling of accomplishment.

Airborne Play

Air and Water Rockets

Air and water rockets are interesting, inexpensive, and relatively safe toys, although children should be warned never to aim them at others or themselves. Park Plastics produces a variety of such projectiles which are sold in toy stores and museum shops for $3 to $9.

Fig. 71

Figs. 71–72. The king of the fuel rocket manufacturers is clearly Estes Industries (Penrose, CO 81240). I wrote to this firm with trepidation: rocketry seemed unreasonably dangerous, and not particularly interesting. I was wrong. The Estes literature and conversations with rocketry hobbyists have thoroughly convinced me that rocketry is an excellent, intellectually exciting activity for the right child.

Estes itself is clearly aware of the dangers of fuel rockets. They have a clearly defined safety code, and any kid who can't be trusted to follow it precisely should not be given a rocket. Moreover, the company has marked its rockets in terms of skill level. No matter how smart your child is, start with Level I and move systematically on from there. Estes also provides excellent documentation on the physics of rocketry, with, for example, booklets on calculating trajectory. An extensive network of rocketry clubs can be found across the U.S. (especially in towns and cities with a good hobby shop), and Estes is pleased to tell you where to find other rocket enthusiasts. Again, I was surprised to find that rocketry isn't just a matter of seeing how high you can get a given projectile—photographic attachments and a variety of launching strategies make for further excitement. Fuel rockets should not be flown by anyone under ten years old, and adult supervision is definitely recommended. Starter sets can be purchased for as little as $14, although the deluxe starter set, which includes an aerial camera, retails for $40. Youth groups often purchase a launcher, which members share in order to save on start-up costs.

The National Association of Rocketry (P.O. Box 178, McLean, VA 22101), an organization for people interested in "paper, balsa, plastic, and other non-metallic materials powered by commercial model rocket motors," has local chapters throughout the U.S. and Canada. *Model Rocketry,* the monthly journal of the organization, includes contest announcements, technical features of rocket designs, lots of ads for ordering model kits, and news of N.A.R. activities and special sales.

Fig. 72

Aeronautics Toys

SKY FULL OF PLANES, a sixteen dollar mini-lab from E.D.I., comes with foam sheets, dowels, glue, balloons, crepe paper, landing gear discs, and more—just about everything you need to make simple jets, sailplanes, propeller-driven planes, and hang gliders. *Note:* Save all the cut-out scraps as you make your flying objects; they are called for later.

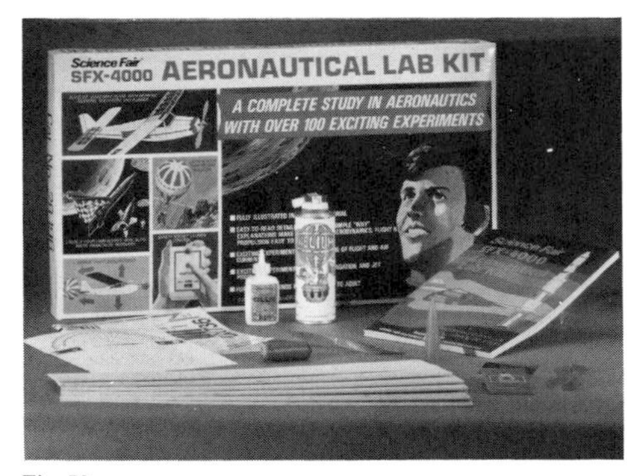

Fig. 73

Fig. 73. An **AERONAUTICAL LAB KIT** from Radio Shack ($20) offers 100 flight projects including a helicopter, model plane, and helium balloon.

Fig. 74. Balsa wood gliders, model kits, and airplanes are inexpensive toys (29 cents on up) which vary in sophistication. The easiest to construct can be completed by a first-grader, while variations on the more difficult models continue to fascinate adults. This World War I vintage **BIPLANE GLIDER** from Paul K. Guillow, Inc., available in most toy and hobby shops, retails for $1. *Note:* These planes do break; think of them the way you think of paper and pencils—resources to be consumed as you learn.

Fig. 75. The **WATER ANEMOMETER,** another Suitcase Science exclusive, is used to measure the force of the wind as well as pump water. A fascinating gadget for children ages six to twelve.

Boomerangs

Fig. 74

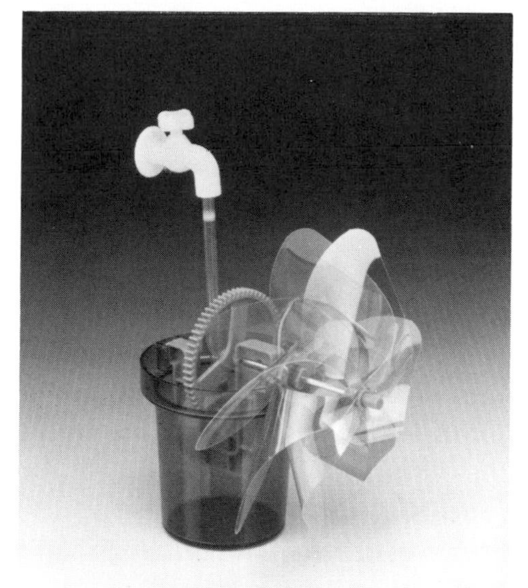

Fig. 75

A boomerang is an object that when thrown correctly returns to the thrower. Although, as in most hobbies/sports, there is a sophisticated variety of styles most experts suggest that kids begin with a NERF or WHAMO boomerang; they're inexpensive and readily available. If the practice and physics of throwing interests you, subscribe to *The Leading Edge,* a $5 newsletter for boomerang fans published bimonthly (Free Throwers' Boomerang Society, 51 Troy Rd., Delaware, OH 43015). It lists tournaments, suppliers, and includes technical tips on construction and flying.

Boomerangs come in a variety of designs and are often beautifully decorated. The following sources sell their pretested versions for $8 to $25; for information write:

Boomerang Man
311 Park Ave.
Monroe, LA 71201

Ben Ruhe
P.O. Box 7324
Washington, DC 20009

Rus-Art
P.O. Box 187
Agawam, MA 01101

Leading Edge Boomerangs
51 Troy Rd.
Delaware, OH 43015

The United States Boomerang Association has its own newsletter, contests, and so on. Write: U.S.B.A., 4030-9 Forest Hill Ave., Richmond, VA 23225.

Kites

Kite flying is another activity that requires at least an intuitive understanding of physics. Experts note that paper kites are much more difficult to fly than those made of plastic, Tyvek, Mylar, or a rip-stop fabric.

If you're interested in kite flying as a hobby, sit down with an issue of *Kite Lines* (4/year, $11, 7106 Campfield Rd., Baltimore, MD 21207), available at newsstands as well as by mail. This is a beautiful publication, chock full of information and ads for the newest and latest in kites.

The national organization of kite enthusiasts is called the American Kitefliers Association. For membership information write: A.K.A., 113 W. Franklin St., Baltimore, MD 21201.

IN PRINT

For Grade-Schoolers

Ardley, Neil, *Making Things Move* (Watts, 1984).
Good experiments with everyday objects, accompanied by excellent explanations. Won high praises from scientists and kids. Other physics books in this series include *Exploring Magnetism, Hot and Cold,* and *Working With Water.*

Bradley, John, *The Plane* (Macmillan, 1985).
An attractive and ingenious pop-up book describing the ins and outs of a Boeing 747 jet.

Bronowski, Jacob, and Millicent Selsam, *Biography of an Atom* (Harper & Row, 1965).
This now-classic work for children in grades 3 to 7 describes the cycle of a carbon atom by including information from astronomy, physics, biology, and chemistry.

Chase, Sara B., *Moving to Win: The Physics of Sports* (Messner, 1977).
The laws of physics are apparent in this novel description of athletic success.

Goor, Ron and Nancy, *Shadows: Here, There, and Everywhere* (Crowell, 1981).
A masterful combination of art and science, this book explains why and how shadows come to be. Suitable for even young children.

Kaufmann, John, *Fly It* (Doubleday, 1980).
Directions for making and improving upon your own kites, boomerangs, helicopters, hang gliders, and hand-launched gliders, using household materials.

Kramer, Anthony, *The Magic of Sound* (Morrow, 1982).
Experiments help children understand more about acoustics in general and hearing in particular. Interesting applications of this information—industrial noise, sound effects, and more—are also highlighted.

Lefkowitz, R. J., *Forces in the Earth: A Book about Gravity and Magnetism* (Parents, 1974).
The clarity with which difficult ideas are explained is impressive indeed. Includes experiments for children to try.

Renner, Al G., *Experimental Fun with the Yo-Yo and Other Science Projects* (Dodd, Mead, 1979).
By manipulating variables kids learn about the physics of yo-yos, cork sailboats, hang gliders, cartesian divers, and more.

Schneider, Herman and Nina, *Science Fun with a Flashlight* (McGraw-Hill, 1975).
Learn about the sun, moon, colors, shadows, and more with this excellent selection of experiments.

Simon, Seymour, *Mirror Magic* (Lothrop, Lee & Shepard, 1980).
Interesting activities explain how and why mirrors work.

Wilkes, Angela, and David Mostyn, *Simple Science* (Usborne/Hayes, 1983).
A challenging and entertaining book of physics experiments. Cartoon-like drawings make directions clear and inviting.

For Advanced Readers

Anderson, Norman D., *Investigating Science in the Swimming Pool and Ocean* (McGraw-Hill, 1978).
One of a series by Anderson which proves that physics toys are available everywhere.

Apfel, Necia, *It's All Relative* (Lothrop, Lee & Shepard, 1981).
"Thought experiments" are used to explain the theory of relativity, the relationship between time and space, gravitation and acceleration. A favorite among physics teachers.

Branley, Franklyn, M., *Color: From Rainbows to Lasers* and *The Electro-Magnetic Spectrum* (Crowell, 1978 and 1979).
The former director of the Hayden Planetarium explains very difficult concepts, their history, and current applications in terms a non-scientist can grasp.

Ito, Toshio, and Hirotsugu Komura, *Kites: The Science and the Wonder* (Harper & Row, 1983).
The best book on kite aerodynamics currently available.

Kent, Amanda, *Physics* (EDC Publishing, 1984).
Experiments on sound, electricity, light, color, magnetism, heat, air, and optics made this one an excellent introduction for good students new to the field.

Ruhe, Ben, *Boomerang* (Minner Press, available through *The Leading Edge,* or Ruhe himself, addresses listed on page 215, 1982).
All the basics are right here.

Zubrowski, Bernie, *Bubbles* (Little, Brown, 1979).
Bubble sculptures, bubble solutions, bubble technique, and more make bubble physics a subject worth your time.

WEATHER

Weather is one of those interdisciplinary studies that involves physics, earth science, and engineering. Because weather is concrete and familiar to us all, it is a good place for children uncomfortable with science to begin experimenting and reading. Because the study of weather involves a complicated array of ideas and techniques, it is also a good topic for those with a "scientific bent."

HANDS-ON

Weather Instruments

Forecasting and measurement tools—thermometers, barometers, and wind meters—combined with a good book on weather forecasting and a weather diary make an excellent gift.

WEATHER CHEK, a nifty $5 weather station which indicates wind direction, wind speed, rainfall, and temperature is available at toy stores and through Nasco.

The **WEATHER FORECASTER,** an out-of-print kit from Skilcraft, is still available in some toy stores and through catalog houses. It is an excellent science kit, designed for age ten and older, which originally sold for about $30.

A **DAILY WEATHER LOG,** $8 from Edmund, helps amateur meteorologists keep track of their data on temperature, humidity, cloud formations, and wind speed.

IN PRINT

For Young Children

Branley, Franklyn M., *Flash, Crash, Rumble & Roll* (Crowell, 1985).
Instead of hiding under the covers in a thunderstorm, take out Branley's excellent explanation of natural phenomena and answer all the obvious questions. To allay fears, try *Thunderstorm* by Mary Szilagyi (Bradbury, 1985).

———, *Rain and Hail* (Crowell, 1983).
Examples from everyday life illuminate the process by which water vapor returns to earth as rain or hail in this clear beginning-to-read book.

Webster, Vera, *Weather Experiments* (Children's, 1982).
Simple experiments for the home meteorologist.

Wolff, Barbara, *Evening Gray, Morning Red: A Handbook of American Weather Wisdom* (Macmillan, 1976).
Rhymes about natural occurrences are followed by factual explanations of the phenomena.

For Grade-Schoolers

Compton, Grant, *What Does a Meteorologist Do?* (Dodd, Mead, 1981).
A good career book for the weather-minded.

Fodor, R. V., *Frozen Earth: Explaining the Ice Ages* (Enslow, 1981).
In explaining the ice age, Fodor relies heavily on astronomical theory.

Sattler, Helen R., *Nature's Weather Forecasters* (Lodestar, 1978).
Nature provides signs that the weather is about to change—a fascinating examination of these indicators.

For Advanced Readers

Ludlum, David, *The Weather Factor* (Houghton Mifflin, 1984).
Ludlum's intriguing compilation of obscure facts about the influence of weather on American history engenders an appreciation of both subjects.

Stommel, Henry and Elizabeth, *Volcano Weather: The Story of 1816, The Year Without a Summer* (Seven Seas, 1983).
An accurate and lucid discussion of how the clouds of ash from the eruption of an Indonesian volcano affected the weather in New England and northern Europe.

Williams, Terry Tempest, and Ted Major, *The Secret Language of Snow* (Sierra Club/Pantheon, 1984).
This lively account of how scientists rely on native people's descriptions of snow illustrates the way language affects perception.

And a Magazine

Weatherwise, 6/year, $22 (4000 Albermarle St., N.W., Washington, DC 20016).

This publication for amateurs includes articles of current and historical interest, as well as a review of the previous month's weather data, and weather experiments. The vocabulary and style is accessible to a talented junior-high student.

17 Astronomy: The Sky's the Limit

Astronomy is the study of celestial bodies, their size, how they move, and their composition. Although most people associate astronomy with telescopes and the night sky, this science also includes the study of the earth in relation to other planets, the activities of the space program in general and the astronauts in particular, and the related science of physics. Amateur astronomers, unlike amateurs in most other fields, have made significant scientific contributions to the field, and are generally welcomed by professionals. In short, amateur astronomy provides a variety of challenges and entry points, and may survive as a rewarding lifelong hobby.

HANDS-ON

To most children, astronomy is virtually synonymous with owning a telescope. So they talk their parents into plunking down $50 or $60, no small sum to be sure, on a nicely packaged item from their local toy supermarket, and set out to explore the night sky. The evening starts out well; the moon is an impressive sight through one of these instruments, and expectations build. Next the child turns to a star, any star. Surprisingly the dot on the sky looks slightly bigger, and considerably blurrier, through the new telescope. At random they try star after star, with no effect.

In fact, the children described here have two problems which make the satisfactions of astronomy inaccessible. The first problem is conceptual: without a roadmap, without some understanding of the differences between planets and stars, without an appreciation of distances, without a consciousness of order, sky watching is boring. The second problem is that these amateurs have purchased a telescope that is simply inadequate. The instrument shakes, it is overpowered, and the resolution is poor.

To address these problems, I recommend first that your child demonstrates some mastery of the sky, and that he/she articulates reasonable ideas about what is expected of the telescope, before you buy one. There are a number of books, magazines, and clubs designed to help children find their way in observing; begin with these. Second, recognize that any adequate telescope you buy new will probably cost $150 or $200. Before blanching, remember that a telescope, unlike a computer, where the technology improves daily, is an investment in an item that remains valuable. Realize also that there are good alternatives to the "instrument intensive" notion of sky viewing.

Star Charts

The stellar equivalent of a road map is the star wheel, a rotating device that pictures the night sky at a given time and date. Unless otherwise stated, the star charts are designed for people at 30° to 50°N. (Check your latitude on an atlas.) The following popular star charts are currently available:

Edmund's rotating STAR AND PLANET INDICATOR is especially easy to use and sells for the bargain price of $2.50. The explanatory material enclosed is also excellent.

The STAR FINDER AND ZODIAC DIAL, which sells for approximately $4 from Nasco and many bookstores, uses a slightly larger format and features a nice luminous dial. The documentation provided is not detailed, and is best supplemented by a book on sky watching.

A large-format book, *Seasonal Star Charts,* from Hubbard Scientific ($8.95), includes a rotating luminous dial, laminated for dewy nights, along with detailed star charts and lists of interesting celestial objects for the telescope and naked eye.

DIAL THE NIGHT SKY, a $6 star chart from Sky Publishing (49 Bay State Rd., Cambridge, MA 02238-1290), is available for latitudes 20° to 32° N, 30° to 40° N, and 38° to 50° N. (Most of the other charts are designed for approximately 30° to 50° N.)

The **SKY CHALLENGER,** a new concept in sky maps from the Lawrence Hall of Science (Berkeley, CA), comes with six interchangeable wheels which present various views of the heavens (for example, one simply shows the constellations, another highlights nebulae and star clusters, and another is a Native American star chart). Available from either the Discovery Corner, Lawrence Hall of Science, or Hubbard Scientific.

A decorative and useful "global equivalent" of the star map, two lucite domes that fit one into the other to form the **BOWL OF NIGHT,** provides a curved view of the astronomical sky. This attractive model, set at 30° to 50° N, clearly shows how the sky changes through the seasons. Available for $65 from Sherical Concepts, Inc., 941 Parkview Dr., King of Prussia, PA 19406.

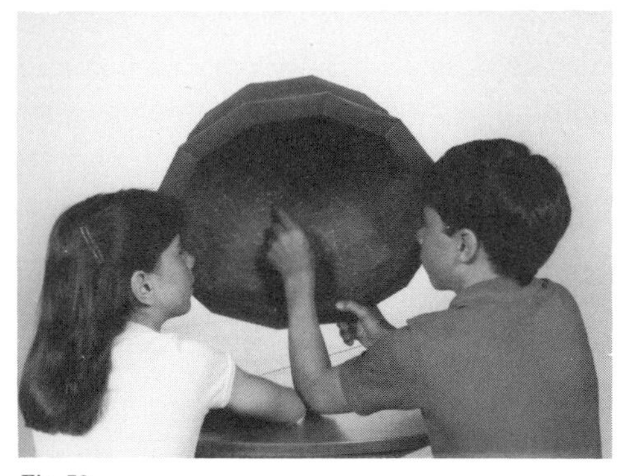

Fig. 76

Fig. 76. The build-it-yourself paper version of this spherical view, known as the **ASTRODOME,** is available from Sunstone (P.O. Box 788, Cooperstown, NY 13326) for $8. Originally designed for West Germany, it features a sky at 50° N, just right for Canadians and Alaskans. Be warned, however, that this model is large, difficult to build, and not easily moved. Recommended for children age twelve and older.

Four times a year *Sky & Telescope* prints a star chart for the southern hemisphere. Monthly charts in this periodical span an exceptionally wide range, 20° to 50° N.

Note: Luminescent star charts can be used in a dark room as a self-quiz on the constellations.

Binoculars

Binoculars are essentially small telescopes. Most people are aware of their use in terrestial observing; however, good-quality binoculars also provide views of the moon and planets much like those Galileo enjoyed through his telescopes. For astronomical observation, experts prefer either 7 × 50 or 10 × 50 with a coated lens (to reduce light loss from glare), although a 7 × 35 set, purchased for terrestial observation (see Chapter 11), can also work well. In addition, weight is a consideration: if the binocular is too heavy a tripod may be required.

Telescopes

What's Available

There are three types of telescopes available today (Fig. 77).

- The oldest design is the refractor, popular telescopes for viewing the moon and planets. They provide sharp, clear images, require little or no maintenance, and take some abuse. Ideally a refractor is attached to a tripod, to stabilize it and make viewing easier. Reaching the eyepiece, however, may become difficult. To compensate, good refractors often include a diagonal mirror, which makes viewing more comfortable. Quality refractors generally cost more than comparable reflectors.
- Reflectors, whose design is credited to Sir Isaac Newton, are the traditional favorite of amateurs; they are generally seen as "good scopes for the price." Because the reflector tube is open, the mirror must be protected from dust and debris. Mirrors can be recoated, but chips or scratches are potentially devastating. Because the image is viewed from an eyepiece at the top of the telescope, users may require a stepstool. In this sense, viewing through a reflector can be awkward, since the telescope has to swing to keep an object in view.
- Catadioptrics are the third group of telescopes currently available for amateurs. These instruments, like the Schmidt-Cassegrian or Muscatov, use a series of mirrors, which actually shorten the requisite length of the tube and make the telescopes more portable. Like the refractor, they are completely closed, relatively rugged, and have an eyepiece at the end of

the tube. The bad news is that the increase in optical surfaces means an increase in price.

In looking for a telescope there are two clues that give you a sense of an item's acceptability. In general, poor telescopes, like toy microscopes, tend to be overpowered. As a rule of thumb, a telescope's *maximum* magnification should not exceed 60 times the value of the aperture in inches. (The aperture is the size of the lens in a refractor, or the size of the mirror in a reflector or catadioptric.) In practical terms, a telescope with a 4″ aperture that claims to magnify more than 240X is producing a distorted and indistinct image. You will be much more satisfied with a scope that offers power 30 to 40 times the aperture—for example, 150X with a 4″ reflector—than with an instrument that offers excessive magnification.

The second clue is the telescope mount. Stars and planets are such tiny images that any vibration in the telescope makes gazing nigh impossible. Furthermore, stars and planets are not fixed, but seem to be continually

TELESCOPES FOR AMATEUR ASTRONOMY

	REFRACTORS	NEWTONIAN REFLECTORS	CATADIOPTRIC TELESCOPES
LIGHT PATH	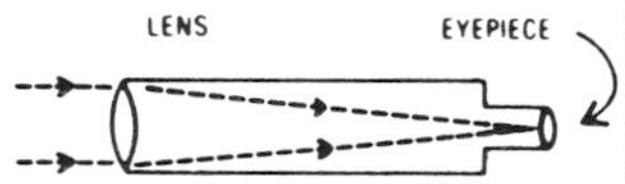 LENS, EYEPIECE MAIN LENS FOCUSES LIGHT	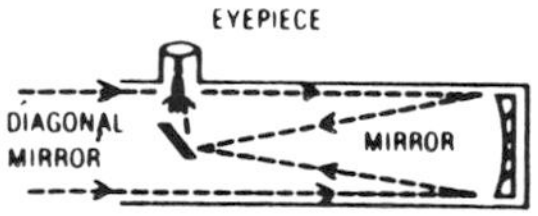 EYEPIECE, DIAGONAL MIRROR, MIRROR PARABOLIC MIRROR COLLECTS LIGHT	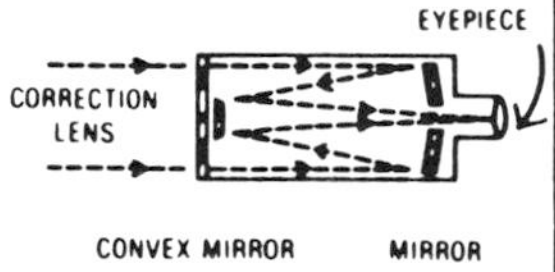 EYEPIECE, CORRECTION LENS, CONVEX MIRROR, MIRROR
APERTURE RANGE IN COMMON USE	2-4 INCHES 60mm (2.4-inch) MOST POPULAR	3-12 INCHES 6-inch MOST POPULAR	3 INCHES and UP
FOCAL RATIOS	f/8-f/17 MOST COMMON	f/5-f/10 f/4 WIDE-FIELD	f/10-f/15

Fig. 77

Fig. 78

moving across the night sky. (Actually the earth moves.) Just as one cranes one's neck to keep a farmhouse in sight while speeding by in a car, so too your telescope needs to swivel with the motion of the stars and planets. In short, you do not want a wiggly, short tripod, or worse yet, a tripod with a clamp that must be loosened each time the telescope is re-aimed. The terms "altazimuthal mount" or "equatorial mount" are generally a good sign. In all cases, an adequate tripod feels sturdy.

Note: Never buy a telescope because it provides a good view of the moon. *Any* telescope can do that.

An exceptionally clear, well-organized, and informative pamphlet for the layperson, *Selecting Your First Telescope,* is available for a donation of $2 from the Astronomical Society of the Pacific, 1290 24th Ave., San Francisco, CA 94122. It includes nontechnical advice on what to look for in a first telescope, a list of reputable telescope manufacturers, and an extensive bibliography of astronomy books.

Fig. 78. The "rich field" or "wide field" telescope, typified by Edmund Scientific's ASTROSCAN, which retails for approximately $300, is designed as an exception to the magnification and tripod rules cited above. Although some feel that the images are not quite as clear as those provided by an excellent traditional telescope, it has three distinct advantages. First, it is smaller, lighter, easier to handle, and therefore more portable than other scopes. Second, because it provides a larger view of the sky, objects stay in sight considerably longer. And third, it provides excellent views of comets and deep-space objects like the Milky Way or star clusters. The ASTROSCAN can be held comfortably in your arms or mounted on its own self-contained base.

Where to Buy

Recently a number of telescope specialty shops have opened across the country which sell via mail order. They often advertise in *Sky & Telescope* or *Astronomy.* Museum stores and catalog houses, especially Edmund, generally carry a good line of instruments. Competition has bred occasional sales; however, the holiday season is not always the time to look for "the good buy."

There is also a market in used telescopes. Again, *Sky & Telescope* and *Astronomy* contain used-telescope ads. If properly maintained, a telescope should last forever.

Building Your Own Telescope

Building a telescope is a time-consuming process that takes patience, precision, and some knowledge of optics. It appears that the people who like to build telescopes are not necessarily the same people who enjoy stargazing. Because of the simplicity of design, a reflector is the telescope to build. Basically, building a telescope means grinding a mirror into a parabola. Without a relatively dust-free workspace where the mirror grinding can continue without disruption, the project, which may last for months and months, is difficult to undertake.

If your youngster (this is not a project for little children) decides to build a telescope, find help from another experienced amateur by contacting your closest astronomy society or local science museum. Several museums even sponsor classes on mirror grinding. Books on telescope making are also available: the classic is *Amateur Telescope Making: Book One,* edited by Albert G. Ingalls (Scientific American, 1974). Mirror-grinding supplies and other good guides are advertised in astronomy magazines such as *Sky & Telescope.*

An Excellent Beginning Astronomy Course . . . By Mail

For adult novices trying to learn astronomy with their children, or for good middle-school students on their own, Edmund produces an excellent program entitled PASSPORT TO THE STARS by Terrence Dickinson. In a carefully designed series of lessons one learns about choosing telescopes, observing double stars, globular clusters, star maps, observer logs, and astrophotography. The $50 price tag is softened considerably by the inclusion of star and planet locators and discount coupons for telescopes and binoculars.

NASA also has several programs designed to promote more in-school interest in space science. For instance, NASA now distributes tomato seeds that have spent a year in space. Children are asked to grow and observe differences between "space seeds" and "earth seeds."

For more sophisticated experimentation, NASA has established the Shuttle Student Involvement Program for Secondary Schools (SSIP-S). Individual students, usually with the assistance of teachers and even outside experts, submit proposals to NASA for experiments to be conducted on the space shuttle. For entry rules, an experiment guide, and information on the space shuttle, write: SSIP-S, National Science Teachers Association, 1742 Connecticut Ave., N.W., Washington, DC 20009.

Games

Fig. 79. Although one can learn the locations of heavenly bodies by simply practicing with star maps, a challenging and rewarding game like **STELLAR 28** produced by Hubbard Scientific ($6.95) can really help. The game, which comes with a deck of cards and a board depicting the night sky, actually includes twenty-eight different games (both for groups and solitaire) which vary in difficulty (highly recommended for age ten to adult).

Fig. 79

YOTTA KNOW STARS AND PLANETS comes with a deck of about sixty cards and a board/grid ($13.95 for cards and boards, $6.95 for cards alone). The object is to correctly identify the pictures of planets, constellations, nebulae, and galaxies, although none of these are located on a sky map. Still, the photographs are attractive and the information provided on the reverse side is helpful and accurate. As in **STELLAR 28,** winning the game is clearly related to astronomical knowledge. A good game for age eight and older.

THE SOLAR SYSTEM, fifty boxed experiments for children in grades 4 to 8 from Educational Insights, includes ideas for experiments, demonstrations, and activities to be carried on independently. Some of the suggestions are classic—sitting on a piano stool and noticing what happens when arms are moved in and out, or making a homemade telescope using two magnifying glasses. Other cards introduce children to facts and ask them to work with the information given. A bit of history and folklore is also included. More akin to a good learning center than a toy or game, this one retails for $6.95 from places like Nasco.

Note: There are a number of "space" board games available that require some strategy, but no knowledge of astronomy. In this sense they are not science-related.

Junior Planetariums

Hubbard sells a small plastic project planetarium for $9.90 which consists of a model of the sun, earth, and moon. A map of planet orbits is also included. With this miniature planetarium a child can understand eclipses and the motions of the planets relative to the earth.

There are at least two junior planetarium sets on the market, both of which shine light through tiny pinpricks on a cantaloupe-size globe to project the night sky on the ceiling and walls of a dark room. The rotating globe is a nice feature, and the star charts and pointer are useful indeed. But even with this piece of equipment, which costs up to $40, learning the sky is largely a matter of practice. The bargain version of the "make your room look like a sky" kit is available from Radio Shack for $12.95. Plastic plates marked with the constellations are slipped onto a lighted dome. The effect is good, the price is right, and the educational benefits are all there.

A set of 250 star decals which glow in the dark—sometimes called **SEEIN' STARS**—can turn a room into a large star chart. Instructions are included, although setting up a room with this many 1″ to 2″ stars takes some patience; $4 from science stores or museum shops.

Solar Activities

The sun, surely our most important star, is a good target for young observers who can be trusted to follow this crucial mandate: *Never look at the sun directly, even with sunglasses or through dark glass.* The sun is viewed by projecting it onto a screen or board. Some telescopes, notably refractors, come with a plate to project the image of the sun through the eyepiece. A sun screen is also available for the ASTROSCAN.

Sundials and solar calendars also provide an entry into solar observation. Several inexpensive versions are available. Hubbard Scientific sells two different sundials and a solar calendar, each for $4.95. All three are made of cardboard, require some construction, and include good explanations. Edmund Scientific sells a sturdy combination sundial and sextant made of plastic and cardboard for $15.95.

Dial-a-time ($5) uses a cardboard wheel to encourage children to set the time in other parts of the world. One hopes that the idea that the time zones of earthlings differ may provoke obvious questions about rotation of the planet, sunrises, sunsets, and so on. The wheel is also nice for a child who is traveling or who is interested in what someone he/she cares for in another part of the world might be doing at this very moment.

ASTRONAUTS AND THE SPACE SHUTTLE

NASA, the acronym for the National Aeronautic and Space Administration, is the government agency involved in sending astronauts into space as well as informing the public about the space program. NASA has several regional centers which provide brochures, photographs, slides, and tapes to interested visitors. In addition, some of the NASA space centers have tours and exhibition halls. Addresses are as follows:

Ames Research Center
Moffet Field, CA 94035

Goddard Space Flight Center
Greenbelt, MD 20771

Lyndon B. Johnson Space Center
Houston, TX 77058

John F. Kennedy Space Center
Kennedy Space Center, FL 32899

Langley Research Center
Langley Station
Hampton, VA 23365

Lewis Research Center
21000 Brookpark Rd.
Cleveland, OH 44135

George C. Marshall Space Flight Center
Marshall Space Flight Center, AL 35812

IN PRINT

Recent space probes have revolutionized our understanding of the planets. For this reason, check the copyright dates of all astronomy books before as-

Inexpensive Gifts for Astronaut Fans

A catalog of space toys, gifts, posters, patches, and T-shirts is available for $1.25 from Action Packets, 344 Cypress Rd., Ocala, FL 32672. Here is but a sample of their offerings:

Freeze-dried astronaut ice cream, $1.25

Patches from space missions, $1 to $2 each

A pen that writes in freezing cold, boiling heat, or zero gravity, $8

Space-shuttle blast-off T-shirts, $5.50

Plastic scale model kits of the shuttle, $3.50 to $15

NASA photograph posters, $1 to $2

suming you have up-to-date information. Astronomy buffs are also fascinated by color photographs, although many excellent space books do not include color pictures because of cost. If a book you like lacks photographs, supplement it with a highly pictorial guide such as T. Ferris's *Galaxies* (Sierra Club Books, 1980), H. Friedman's *The Amazing Universe* (National Geographic, 1975), or the more technical *The New Solar System* by J. Kelly Beatty, Brian O'Leary, and Andrew Chaikin (Sky Publishing Company, 1982).

For All Ages

Allen, Joseph P., *Entering Space: An Astronaut's Odyssey* (Workman, 1984).
The 200 color photos and diary-like text make this space traveler's voyage more real and astounding than those that come before.

Joels, Kerry M., and Gregory P. Kennedy, *Space Shuttle Operator's Manual* (Ballantine, 1982).
Covers all aspects of space flight including eating and emergency procedures. The perfect book to take to the Air and Space Museum in Washington, DC.

Ottewell, Guy, *To Know the Stars: A Simple Guide to the Night Sky* (Astronomical Calendar Pub., Physics Dept., Furman University, Greensville, SC, $5).
These simplified maps of the night sky powerfully reveal the complexity of the heavens.

For Young Children

Blocksma, Mary and Dewey, *Easy-to-Make Spaceships That Really Fly* (Prentice-Hall, 1983).
Paper plates and Styrofoam cups are made airborne through the simple directions and diagrams presented here.

Branley, Franklyn, *Eclipse: Darkness in Daytime* (Crowell, 1973).
Both a history and an explanation, perfect preparation for the great event.

———, *The Sky Is Full of Stars* (Crowell, 1981).
An excellent sourcebook to help young children locate stellar objects.

Fields, Alice, *The Sun* (Watts, 1980).
An informative guide to the size and composition of the sun, and its effect on earth. Includes simple experiments.

Gibbons, Gail, *Sun Up, Sun Down* (Harcourt Brace Javonovich, 1983).
A little girl explains the power of the sun and how it is responsible for seasonal changes, warmth, power, plant life, and more in this boldly illustrated picture book.

Livingston, Myra Cohn, *Sky Songs* (Holiday, 1983).
Livingston's poems are paired with superb paintings by Leonard Everett Fisher to reveal the sky through all its changes: temporal, seasonal, and climatic.

Simon, Seymour, *Earth: Our Planet in Space* (Four Winds, 1984).
The long view from an author whose science and style are very accessible.

———, *Jupiter* (Morrow, 1985) and *Saturn* (Morrow, 1985).
A look at two planets from our own galaxy, as appealing to older children with reading difficulties as to the younger readers for whom it's intended. These are the newest in Simon's series on our solar system.

For Grade-Schoolers

Ford, Adam, *Spaceship Earth* (Lothrop, Lee & Shepard, 1981).
One of the few books for children of this age which discusses light, gravity, and the Big Bang, as well as meteors, planets, and asteroids.

Gallant, Roy, *101 Questions and Answers About the Universe* (Macmillan, 1984).

This book, based on questions actually asked by young planetarium visitors, should appeal to field-trippers and space enthusiasts alike.

Jabber, William, *Exploring the Sun* (Messner, 1980).
The history of solar interest, from earliest worship to modern studies.

Rey, H. A., *Find the Constellations* (Houghton Mifflin, 1976).
A classic young person's guide to the night skies, by the knowledgeable author of the Curious George series.

Simon, Seymour, *How to Be a Space Scientist in Your Own Home* (Lippincott, 1982).
Experiments to demonstrate gravitational effects, rid the air of carbon dioxide, measure acceleration, and send messages to extraterrestrial life-forms.

Snowden, Shelia, *The Young Astronomer* (Usborne/Hayes, 1983).
This highly pictorial guide to what amateur astronomers see in the night sky through a small telescope includes practical hints and uncommon facts.

Spooner, Maggie, *Sunpower Experiments: Solar Energy Explained* (Sterling, 1980).
The principles of solar energy explained through hands-on activities.

Weiss, Malcolm, *Far Out Factories: Manufacturing in Space* (Lodestar, 1984).
Potential applications of space technology in manufacturing, with a discussion of military applications.

Yount, Lisa, *The Telescope* (Walker, 1983).
Part of a lively and fascinating series on "inventions that have changed our lives," this one focuses on optical instrumentation.

For Middle-Schoolers

Apfel, Necia H., *Astronomy and Planetology* (Watts, 1983).
Assorted space projects—outlines for some, detailed descriptions for others—with a listing of competitions to enter.

Asimov, Isaac, *How Did We Find Out About the Atmosphere?* (Walker, 1985).

From the earliest experiments proving the existence of air to the relatively recent discoveries of rare gases, a masterful science writer tells us why we (and seven other planets in our solar system) have an atmosphere.

Dickinson, Terence, and Sam Brown, *The Edmund Sky Guide* (Edmund Scientific Company, 1977).

The $5 price tag makes this excellent introduction to the sky for telescope owners a "best buy." Includes excellent sky charts.

Edmund MAG 5 Star Atlas or *Edmund MAG 6 Star Atlas* (frequently revised).

A good bet for the amateur who has "learned the basics," these two inexpensive paperbacks contain detailed maps, additional lists of interesting objects to be viewed by telescope, and practical hints on how to observe. *Note:* Although sixth-magnitude are fainter than fifth-magnitude stars, both are within reach of most telescopes.

Gallant, Roy A., *Once Around the Galaxy* (Watts, 1983).

Gallant's history of astronomy is a case study in the way scientific theories have changed and are still open to revision.

Herbst, Judith, *Sky Above and Worlds Beyond* (Atheneum, 1983).

An enthusiastic journey through the universe—humorous, accurate, and inviting.

Lauber, Patricia, *Journey to the Planets* (Crown, 1982).

The prominent features of each planet in our solar system are accurately explored by a masterful science writer.

Moulton, Robert R., *First to Fly* (Lerner, 1983).

A description of the ideas and testing procedures that took student Todd Nelson's project from a proposal to an experiment aboard the third shuttle mission.

Muirden, James, *Astronomy with Binoculars* (Arco, 1984).

Excellent practical advice from a distinguished amateur astronomer.

———, *Astronomy Handbook* (Arco, 1982).
How to use a telescope, locate stars, and photograph the sky. A knowledgeable amateur shares his enthusiasm with readers.

Neely, Henry M., *The Stars by Clock & Fist* (Viking, 1972).
Using your outstretched fist as a measuring device, and setting north to 12 o'clock, the author shows how you can find your way around the night sky.

Simon, Seymour, *Long Journey from Space* (Crown, 1982).
Just in time for the return of Halley's Comet, Simon describes meteors and comets.

Snowden, S., *The Young Astronomer* (EDC Publishing, 1983).

Vogt, Gregory, *The Space Shuttle* (Watts, 1983).
All about two programs for high-school students interested in sending their own experiment into space.

Whitney, Charles, *Whitney's Star Finder* (Knopf, 1985).
Perhaps the best known of the star guides, Whitney's excellent book describes popular stellar objects and offers practical information for the novice. Sunrises and sunsets, eclipses, meteors, observing the planets with binoculars, stars, and clusters are discussed. Includes useful appendixes (such as a list of meteor showers) plus a star-finder wheel.

Magazines

Odyssey, 12/year, $15 (AstroMedia Corporation, 625 E. St. Paul Ave., P.O. Box 92788, Milwaukee, WI 53202).
An excellent magazine of astronomy and outer space for eight- to fourteen-year-olds. The editors have a good sense of children's questions; one feature may describe what and how astronauts eat in space, while others use spectacular color photos to illustrate hard-to-conceptualize information. Included is a monthly star chart, lots of "how to's" (can you use your watch and the sun as a compass?), ads for books and posters, and a well-written question-and-answer column for kids.

Astronomy, 12/year, $21 (AstroMedia Corporation, 625 E. St. Paul Ave., P.O. Box 92788, Milwaukee, WI 53202).

Space Lobbyists

Just as there is a network of amateur astronomers, there are networks of space enthusiasts dedicated to convincing the public and Congress of the advisability of supporting further activities in space.

The L-5 Society, 1060 E. Elm, Tucson, AZ 85719. A group for people who think about space stations, lunar cities, and "Star Wars" defense, the L-5 Society lobbies Congress for more manned flights and produces a glossy monthly magazine for members.

National Space Institute, West Wing, Suite 203, 600 Maryland Ave., S.W., Washington, DC 20024. A well-organized group that lobbies for manned and unmanned space flights. Publishes a monthly nontechnical magazine, *Space World,* and supports educational projects.

The Planetary Society, 110 S. Euclid Ave., Pasadena, CA 91001. Formed by a number of well-known astronomers, including Carl Sagan, this group advocates increased support for planetary exploration by unmanned probes and funding for SETI (Search for Extra Terrestial Intelligence) projects. Publishes an excellent bimonthly nontechnical magazine and supports educational projects.

Radio Amateur Satellite Corporation, P.O. Box 27, Washington, DC 20044. They've built the amazing OSCAR satellites NASA launched for use by ham radio operators.

The magazine to buy for amateurs who have outgrown *Odyssey* but aren't yet ready for *Sky & Telescope.* Feature articles provide good coverage of the shuttle flights with lots of full-color photographs. The magazine includes a monthly star chart and plenty of ads for astronomical supplies, books, and posters.

Sky & Telescope, 12/year, $18 (Sky Publishing Corp., 49 Bay State Rd., Cambridge, MA 02238-1290).

Features general information articles on astronomy written by professional astronomers, a monthly sky chart, a column for telescope builders (both advanced and beginning), updates on the latest comet or solar eclipse, and ads for books, telescopes, and astronomical goodies. This excellent magazine, available in most libraries, requires a background in science.

Amateur Astronomy Organizations

Fortunately, amateur astronomers can be located through one of several organizations. Joining one of these groups, if only for a specific outing, is highly recommended.

Astronomical Society of the Pacific (1290 24th Ave., San Francisco, CA 94122). Brings together professional and amateur astronomers, provides a bimonthly magazine called *Mercury,* a monthly sky chart, an astronomy services directory of planetariums, local amateur astronomy groups, listings of astronomy courses open to the public, and stores that carry astronomy equipment. The society also sells books, posters, photographs, computer software, and novelty items.

The Astronomical League (c/o Don Archer, P.O. Box 12821, Tucson, AZ 85732). This federation of amateur astronomical societies puts you in touch with local chapters. The league also publishes a quarterly newsletter, *Reflector,* which lists events and provides discounts on books. These experts can be extremely helpful in all facets of observing and instrumentation.

Western Amateur Astronomers (A. McDermott, Secretary, P.O. Box 2316, Palm Desert, CA 92261). A group for amateurs on the West Coast.

The Royal Astronomical Society of Canada (124 Merton St., Toronto, Ontario M4S 2Z2, Canada). National organization of professional and amateur astronomers in Canada.

Note: Many of the major science and technology centers have astronomy clubs.

Posters Dramatic 16″ × 20″ reprints of NASA photographs from planetary and manned missions are available from the publishers of *Sky & Telescope* as part of their Spotlight Print Series. A catalog, "Scanning the Skies," which lists astronomy posters, trade books, and atlases, is available free from Sky Publishing Corp., publishers of this periodical (see address under "Magazines," above).

AstroMedia, publisher of *Astronomy,* also sells posters and space art slides.

The Astronomical Society of the Pacific, one of the largest of the astronomy groups, sells astronomical posters and color prints. Write for their catalog (see "Amateur Astronomy Organizations").

Edmund Scientific also sells astronomical posters and charts in their catalog. Particularly recommended is their large, detailed moon map ($3), extremely useful in identifying lunar craters or mountains.

Many of the planetariums stock posters as well as books. The Hansen Planetarium's poster on the solar system, available in many other museum shops, presents an amazing amount of information clearly and attractively.

The most artistic four-color poster of the constellations (the stars glow in the dark) is available from Celestial Arts, 231 Adrian Rd., Millbrae, CA 94030 or The Nature Company. The 35″ × 35″ map sells for $13.

Many of these posters are suitable for framing.

CAMPS

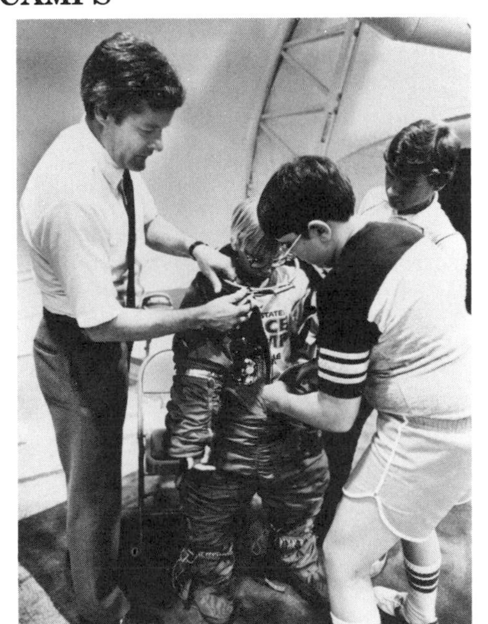

Fig. 80

Midwestern Astronomy Camp (University of Kansas, 1082 Malott Hall, University of Kansas, Lawrence, KS 66045) offers two weeks of instruction for 9th- to 12th-graders in observational and theoretical astronomy.

Fig. 80. A team leader zips a space camper inside a spacesuit during Astronaut Training Day at the Space and Rocket Center Youth Science Program in Huntsville, Alabama. These one-week summer sessions are designed for space enthusiasts who have completed grades 6, 7 or 8. For further information write: Alabama Space and Rocket Center, Tranquility Base, Huntsville, AL 35807.

18 Wired for Electronics

Electronics . . . electrics . . . electricity . . . electrocuted. Since my children were babies I have told them not to play with electricity. This advice is based on some vague notion that metal and water and current can combine to fry children. Let's join Evan, an electronics professional, for a question-and-answer session to discuss my own and similar fears.

WENDY: Hey, Evan, didn't your mother ever tell you not to play with electricity? How did you get started?

EVAN: You seem to have an idea that electronics play means giving your kid a wire and telling him or her to try pushing it into a wall socket. Nothing could be further from the truth. House current can be extremely dangerous, and I don't recommend that any beginner use it as a power source.

My own interest in electronics began when, as an eight- or nine-year-old leafing through an encyclopedia, I found directions for building a crystal radio. By wrapping wire around an empty oatmeal box, and connecting it with a diode and a tuning capacitor, I was able to pick up radio stations in nearby Chicago.

WENDY: But that isn't electronics, is it? What was your power source?

EVAN: Electronics comes from the word electron; electrons are unbelievably small particles. Electromagnetic waves, waves set up by electrons, can be carried along wire to your electric outlets, but they can also pass through air. In picking up a radio station on a crystal radio, you are receiving emissions from a radio tower. Because it is cumbersome to talk about electrons, just as it would be difficult to talk about how many droplets of water fall in a rainstorm, we have found a larger measure called an ampere, or amp for short, for describing the flow of electric current. One amp means 62,800,000,000,000,000,000,000,000 electrons pass a given point each second.

WENDY: What's the difference between an amp and a volt?

EVAN: An amp is the number of electrons that can be delivered from a particular source. Voltage is the power or the pressure with which that power is delivered. But if you are looking for a measure of how much of what to avoid, you also need to know about current. Current, in electronics, is in many ways similar to water current. If you opened up a dam, the person standing at the opening would be crushed by the water's tremendous force. In this sense, water is scary stuff. But just because you know that water can be powerful does not make you afraid to turn on the tap or flush the toilet.

In important respects, the analogy between electronics and water works well. Imagine that you have a water tank. There are several variables that determine the amount of water flowing from that tank. If the tank is completely closed, the water it holds is simply not released. Similarly, in electrical circuits, if no wire is attached to the power source, no electricity is discharged. But imagine that you puncture the water tank and attach a hose to it. If the hose is 1″ in diameter, less water will pour out than if the hose were 8″ wide. Resistors, in electronics, serve this same function; using different resistors is like changing the diameter of the pipe leading out of the tank.

But there is another variable that determines the current of the water released. A 100-gallon water tank with a 1″ hose standing 2 feet off the ground will not exert as much pressure as the same tank with the same hose 50 feet off the gound. In other words, increasing the height of the water tower increases the potential force. In electronics, people use the word "voltage" to describe this phenomenon. The most important formula in electronics, Ohm's law, describes the current available by mathematically relating the amount of voltage to the amount of resistance.

WENDY: How much current is generated from the electronics sets that I've asked you to review? Are they dangerous?

EVAN: All the electronics sets we review use batteries as a power source. A standard dry cell is capable of delivering 1.5 volts. By comparison, a wall plug produces 115 to 120 volts.

A Few Words About Safety

WENDY: For those of us who are particularly concerned with issues of safety, do you have any special words of caution regarding electronics play?

EVAN: The sets you recommend are all very safe as long as one uses a common carbon zinc or alkaline/manganese battery. I do warn children against using nickel/cadmium rechargeable, mercury, and lithium cells for two reasons. First, the materials used in them are highly toxic, and second, if these batteries discharge accidentally, a child could really be burned. (I once had a nickel cadmium battery in my pocket, next to a set of keys. As I reached in, my wedding ring completed the circuit and melted a bit of the ring, causing a painful burn.)

My other safety-related advice is not really aimed at kit users, but at kids who like to tinker.

1. Avoid large capacitors, devices that store and discharge a lot of electrical power.
2. Never work with anything that is plugged into a wall.
3. Never mess with tube devices. It is fine to take apart a battery-powered transistor radio, but an old tube model should be avoided. More dangerous still is a television. Not only does the picture tube store a significant amount of energy which can be inadvertently discharged, the tube itself can implode and cause serious injury.
4. Avoid working with wet hands.

I hope that this conversation has convinced you that the electronics kits listed here are at least as safe as any of the science sets included in this book.

THE PLEASURES OF ELECTRONICS SETS

- Electronics sets can be appreciated on a variety of levels. At one point a child simply enjoys mastering directions and producing a working product, while at other moments he/she focuses on the application of theory by manipulating variables.
- The sets themselves are generally self-contained. You don't need to find additional items (except batteries) to complete your experiments.
- By the time a given project is completed, even novices have garnered a new vocabulary and a new set of manipulative skills.
- Electronics allows you to make something useful, or at least workable—a radio, a burglar alarm.
- Items used for one project can be turned into something entirely new. In this sense, electronics play is versatile and efficient.
- There is immediate feedback about the results of your experiment. It either works, or it doesn't. If you try to fix a nonworking project, your corrective efforts are obviously successful or unsuccessful.
- With kits that include a meter, it is possible to make a simple calculation and to predict what you will measure before you measure it.
- Instead of being told that a theory "just works," you develop a visceral understanding of an idea.

HANDS-ON

Electronics Kits

Fig. 81. E.D.I. produces three electronics **MINI-LABS:** a crystal radio set, an electric bell, and an electric motor with motorboat, each of which sells for approximately $7, and an **ELETROLAB,** which contains all three experiments

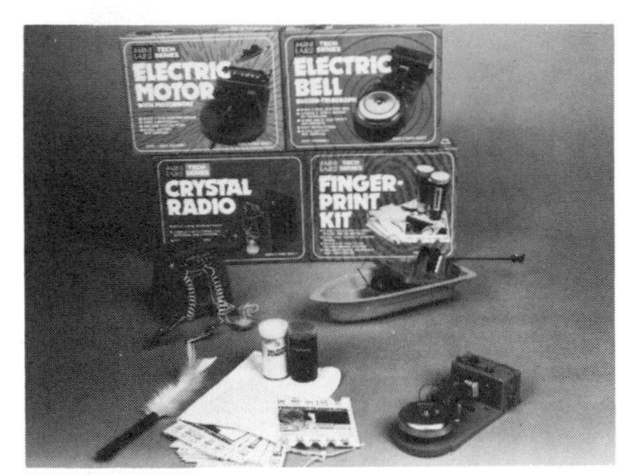

Fig. 81

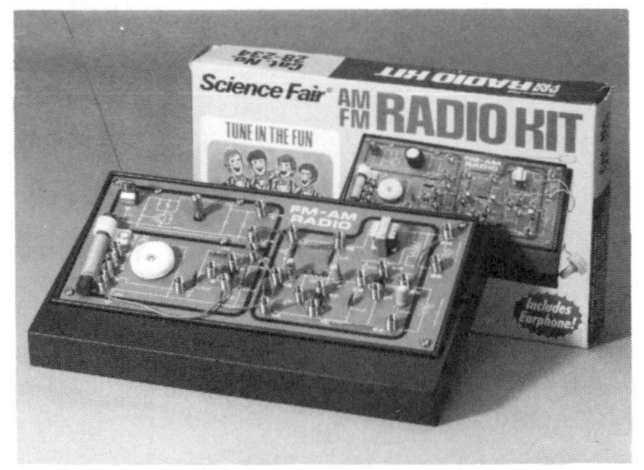

Fig. 82

and retails for $19. The crystal radio, designed for children age nine and older, is the easiest of the three projects to build. In a city one can tune in radio stations, although my rural reviewers, a hearty lot, were fairly impressed that they could make the static softer and louder by using different antennae.

The doorbell and the motorboat each took over an hour to build. The motorboat, for instance, was put together incorrectly by an extremely bright, careful, and dexterous eleven-year-old. He had wrapped the wire incorrectly and not made contact at all points where he should have. In one case the directions were unclear, and in the other, the child used the diagrams instead of written directions and missed an important point. These kits are much more likely to work if a knowledgeable adult helps the child. The explanations of what happens in a given experiment, and suggestions for further experimentation, however, are good.

If you like these junior sets, E.D.I. also produces a fairly elaborate kit called the WORLD OF RADIO AND ELECTRONICS, which retails for $32.

Fig. 82. Radio Shack produces a series of radio kits for young people: an "OLD TIME" crystal radio kit for $5, which like the E.D.I. set uses no tubes, transistors, or batteries; a transistor AM radio for $7; an AM radio broadcasting kit for $8; and an AM/FM RADIO KIT for $13, pictured here. Testers en-

Fig. 83

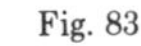

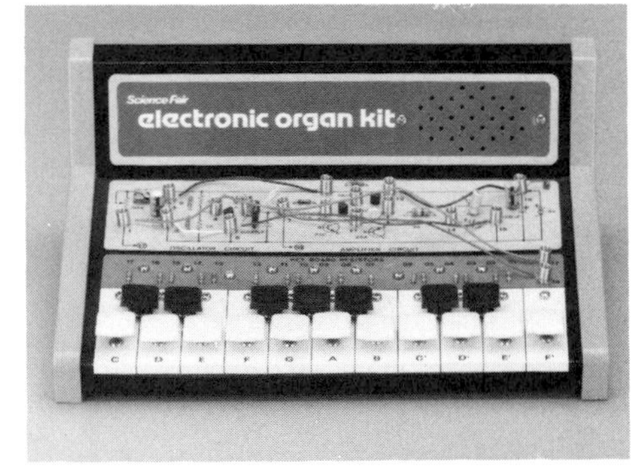

Fig. 84

joyed working with all these sets, found directions clear (if not always easy), and were pleased with their results, which "really worked."

Fig. 83. Radio Shack sells a series of fine electronics toys called the SCIENCE FAIR line. The 10 IN ONE ($10), 30 IN ONE ($15), 50 IN ONE ($20), 160 IN ONE ($30) and 200 IN ONE ($45) all use a system of spring clips—the child simply flips back a spring, tucks in the wire, and the electrical connection is made without any soldering. Each set provides a variety of experiments, and the number of experiments advertised is not inflated. The SCIENCE FAIR line is unique in that the experiments in each set are completely different—one could buy the 10 IN ONE, the 30 IN ONE, and the 160 IN ONE, for instance, for the same child. The experiments do, however, become more difficult as the number of projects increases. The first two sets provide a nice introduction for newcomers, and if the child wants more like it, a simple and satisfying direction is offered.

Fig. 85

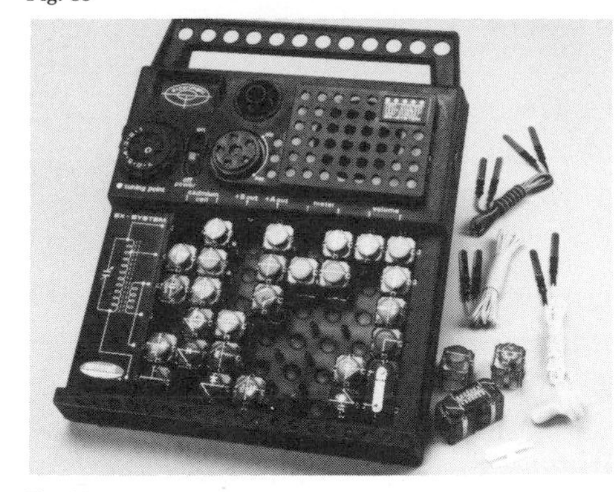
Fig. 86

Fig. 84. An ELECTRONIC ORGAN from Radio Shack ($13) provided a nice change for experienced radio builders. The musical quality of this organ was less than excellent, although all tunes were easily recognized. In this, as in many other electronics sets, when something didn't work the user could go back, step by step, and check for errors.

Fig. 85. N.S.I. imports two excellent Salter Electronics kits from Scotland, available through Toys To Grow On, other catalog houses, and toy stores: the ELECTRONICS 1 kit retails for about $20, and the ELECTRONICS 2, used in this evaluation, for approximately $38. Again, the spring clips used in this series for children age ten and up make connections easy, and all the projects tried really worked. Moreover, the electronics experts commented on the detail and sophistication of the explanations provided. Johnny Ball's tape-recorded message to young users makes this kit a winner. A junior electronics set for about $10 is also available.

Fig. 86. Skilcraft's ELECTRONICS EXPERIMENT SETS (60-project set for $45, 100-project set for $55, and 150-project set for $65) are truly unique. Plastic

Fig. 87

cubes with built-in transistors, diodes, etc., are placed in a grid; using these kits a kid with little dexterity but a sophisticated understanding of circuits can build a number of fascinating projects quickly and easily. Our child testers gave this one four stars—they liked the projects and the design of the kit. Parents also liked the fact that it is neat and portable.

A. P. Products, Inc. (9450 Pineneedle Dr., Box 603, Mentor, OH 44060) produces a series of very inexpensive electronics kits called **HOBBY BLOX**. Each comes with a small plastic "breadboard" for solderless plug-in circuits and directions for "10 Easy-to-Build Projects"; users supply resistors, capacitors, wires, and so on. **HOBBY BLOX** is the most economical way for older enthusiasts to begin, but for children under twelve, the size of the "breadboard" and the style of the directions is probably too difficult.

Fig. 87. Suitcase Science produces several small kits for children age seven to twelve which require the same kind of dexterity and direction-following abilities demanded by the electronics toys. The **ZIG-ZAG MACHINE** and **ELECTRO-MAGNET CAR** ($6 each) are two fascinating, if not altogether sturdy, examples.

Robotics

Electronics fans will delight in twelve different **MOVIT** robot kits from O.W.I. Each creation, which takes a ten-year-old or older approximately 4 hours to assemble, is triggered by sensory input: one responds to noise, another follows along any dark line, yet another veers away from objects in its path. The robots, ranging in price from $25 to $75, require only a small Phillips-head screwdriver, long-nosed pliers, and scissors for assembly, although tweezers for grabbing and an egg carton to keep track of multitudinous screws are also helpful.

Fig. 88. The **MEDUSA,** pictured here, houses a condenser-microphone in its clear plastic dome; an appealing gift for under $30. *Note:* Joseph Electronics (8830 Milwaukee Ave., Niles, IL 60648) sends a free catalog which includes the **MOVIT** series; they mail nationwide. For the name of the distributor nearest you, write directly to O.W.I. (1160 Mahalo Pl., Compton, CA 90220).

Electrics Sets

While electronics kits focus on projects and building, electrics kits are more concerned with teaching the physical principles involved in electronics. Moreover, while electronics kits rely on transistors and diodes, electrics kits focus on safe demonstrations of electricity.

Fig. 89. A JUNIOR ELECTRICS SET from N.S.I. for children eight and older includes screws, wires, bulb and bulb holder, a screwdriver—the basic equipment to build a Morse code flasher, galvanometer, rheostat, and more.

Fig. 90. The ELECTRO-TECH KIT, designed for children age ten and up, contains experiments in electrochemistry, static electricity, and electrical engineering. Users are invited to build a strobe light, burglar alarm, electroscope, and more. (Approximately $25 from N.S.I.)

A POWERTECH ELECTRICS KIT from E.D.I. which retails for about $30 includes a well-organized booklet of experiments on static electricity, magnetism, electrochemistry, and electronics, as well as the pieces required to carry out these experiments. For child testers, especially the project-oriented ones, the electrics set looked like a box of disorganized and unrecognizable "stuff," and they didn't even want to try sorting through it to experiment. This set can, however, answer important questions for kids who want to understand "why."

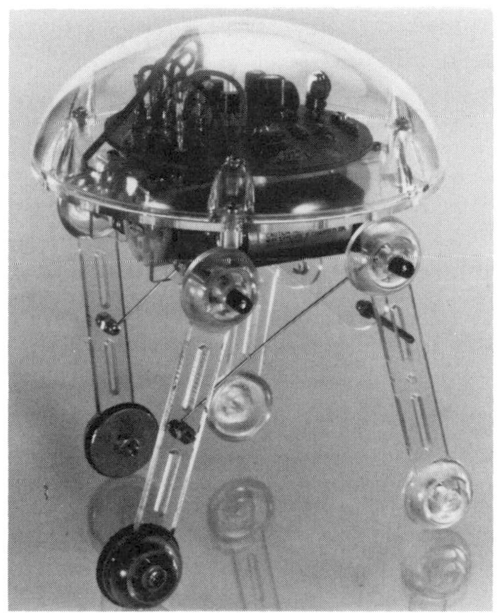

Fig. 88

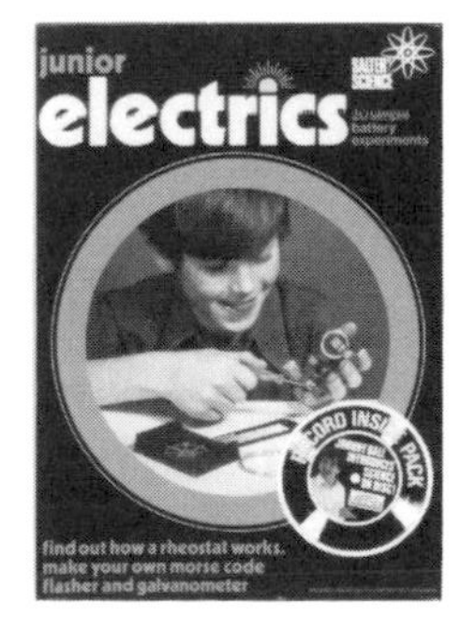

Fig. 89

Fig. 90

Alternatives

Young people who love electronics soon exhaust the projects available in the kits cited above and hunger for more. Users should realize that in working with the pre-fab sets they have learned a vocabulary and technique which allows them to go to most of the project books listed below. Choose an activity, buy the requisite parts, and simply proceed. In this sense, many of the in-print suggestions could well be seen as hands-on.

IN PRINT

It is indeed difficult for anyone looking through a list of books and magazines to know what is appropriately challenging. Electronics experts tell me that a kid who has worked with one of the advanced kits listed above is ready to embark on most of the "magazine-type" projects, although soldering is certainly a useful skill they may wish to learn first. (Heathkit is meticulous in their explanation of soldering, and crystal clear in their labeling of what one needs to know before beginning any given project.)

Most of the electronics magazines and the annual project books they published are available at newsstands. What I recommend, however, is that you look up "electronics parts" in the Yellow Pages of the phone book in your nearest city, and stop by to browse through the holdings at one of these establishments. Knowledgeable people can help you choose an appropriate book or project, and make sure that you have all you need to begin. Another option is to use a service like that provided by Heathkit: projects come complete with materials and excellent directions.

For Younger Readers

Ardley, Neil, *Discovering Electricity* (Watts, 1984).

An introductory experiment book; the early sections on basic properties of electricity are very good. *Note:* The drawings of batteries look remarkably like car batteries; although the author does state that these should be 3 volt, a mistake could be tragic. In addition, the "tongue test" on page 14 should only be done with utmost care.

Wade, Harlan, *A Book about Electricity* (Raintree, 1977).
Part of a series for very young readers, this volume on electricity connects abstract concepts to everyday experience.

For Advanced Readers

Englebardt, Stanley L., *Miracle Chip* (Lothrop, Lee & Shepard, 1979).
Chips have radically changed the entire field of electronics, and this book tells you how.

Graf, Rudolf F., *Safe and Simple Electrical Experiments* (Dover, 1973).
Describes thirty-eight experiments on static electricity, thirty-two on magnetism, and thirty-one on currents and electromagnetism; a nice introduction to the field.

Janeczko, Paul B., *Loads of Codes and Secret Ciphers* (Macmillan, 1984).
Although some of the codes rely on electronic transmission, the basic idea of coding, and the challenge of putting together and taking apart messages, will appeal to kids in grades 4 and above.

Math, Irwin, *Morse, Marconi, and You* (Scribner, 1979).
A nice combination of theory and hands-on experiments for assembling communication devices.

———, *Wires and Watts* (Scribner, 1981).
Hands-on experiments to better understand electricity and magnetism.

Sootin, Harry, *Experiments with Electric Currents* (Norton, 1969).
An explanation of fundamental ideas, rather than a series of magic tricks, this book is useful.

Project Handbooks and Annuals

Electronic Experimenter's Handbook (P.O. Box 2907, Clinton, IA 52735).
This inexpensive collection of projects from *Popular Electronics* is published annually. Reprint series of other projects are available at the same address.

101 Electronic Projects: All Easy to Build ($2, Davis Publishing Inc., 380

Lexington Ave., New York, NY 10017). Electronic games, computer gizmos, and numerous other forms of high-tech gadgetry; this publication is available at newsstands nationwide.

Engineer's Notebook (Radio Shack Technical Publications, Dept. D.G., 1100 One Tandy Center, Fort Worth, TX 76102). This soft-cover publication is written by Forrest M. Mimms III, author of "Experimenter's Corner" and "Project of the Month" in *Popular Electronics,* now called *Computers and Electronics* magazine. Since some of the projects are simply adapted from this source, read it to check the level of difficulty before dipping into this. *Note:* This is just one of many Radio Shack publications which detail projects—see others in your nearest outlet store.

Integrated Circuits Projects (Howard W. Sams & Co., 4300 W. 62nd St., Indianapolis, IN 46268). This company publishes an extensive series of project manuals.

Modern Electronic Circuits Reference Manual (McGraw-Hill, 1980). Written by John Markus, this $75 book is available in most libraries. It includes over 3600 circuits, complete with values of all parts and performance details. Good for quick reference and leisurely browsing.

Magazines

The following magazines are written for adult electronics amateurs, although many sophisticated child users can follow them.

Computers & Electronics, 12/year, $16 (Ziff-Davis Publishing Co., One Park Ave., New York, NY 10016).

Hands-On-Electronics, 4/year, $10 (Gernsback Publications, Inc., 200 Park Ave. South, New York, NY 10003).

Radio-Electronics, 12/year, $13 (Gernsback Publications, Inc., see above).

Mail-Order Catalogs

Heathkit (Heath Company, Benton Harbor, MI 49022).

OK Industries Inc. (3455 Conner St., Bronx, NY 10475). Electronic hobby kits which *do* require soldering, from $7 to $13.

Note: Local electronics houses also print catalogs regularly. Check at your nearest outlet.

Miscellaneous

Directions for hands-on "Science Activities in Energy" are available free from the U.S. Department of Energy, Technical Information Center, P.O. Box 62, Oak Ridge, TN 37830. The set entitled "Electrical Energy" is particularly appropriate for grade-school children trying to understand more about electricity.

AMATEUR RADIO

Amateur radio may be the best organized technological hobby in America. Licensed operators, who range in age from 6 to 106, communicate with people the world over. They are called upon for help in case of disasters, engage in competitions, help monitor races and games, and generally have a wonderful time sharing information with each other or the public. If you are interested in the hobby, write to the American Radio Relay League (225 Main Street, Newington, CT 06111). They will send you free of charge the name of a mentor (they call this person an Elmer) who contacts you, offers advice on or lends you beginning equipment, and guides you through your licensing exams. If you wish to learn more about amateur radio, I heartily recommend the A.A.R.L. beginner packet for $8.50; this includes a cassette tape, narrated by Jean Shepherd, designed to acquaint you with Morse code, and the comprehensive and comprehendable book, *Tune In the World of Ham Radio.* In addition, the A.A.R.L. has established a special rate for young users—kids under 13, who are themselves the youngest radio amateur in the household, can become a member for only $6.25 annually. This entitles them to all membership benefits as well as the monthly publication, *QST*. I am genuinely impressed by the network amateur radio operators have established, the services they provide, and the competence of their operators and operation.

Schwartz, Martin, *Amateur Radio Theory Course* (Ameco Publishing Corp., 275 Hillside Ave., Williston Park, NY 11596, 1977).

A text on radio theory and regulations.

The Radio Amateur's Handbook (The American Radio Relay League, 1986).

A reference tool which helps those who pass their basic license get on to bigger and better things.

The Radio Amateur's License Manuals (American Radio Relay League, 1986).

Describe what you need to learn to upgrade your license.

The FCC Rule Book (American Radio Relay League, updated every four months).

This manual, to be used in conjunction with the license manual series, describes and explains rules as well as quotes them directly.

Tune In the World of Ham Radio (The American Radio Relay League, 1984).

A publication that makes beginning inviting and altogether possible.

Magazines

CQ: The Radio Amateur's Journal, 76 N. Broadway, Hicksville, NY 11801, 516-681-2922. Novices and experts alike give this one high praise.

QST: Devoted Entirely to Amateur Radio. The official magazine of the American Radio Relay League. Write: 225 Main St., Newington, CT 06111.

19 A Computer Primer

The computer industry is changing so rapidly that any specific book information on what machines or software to buy is probably outdated by the time you read it. I use this chapter, instead, to talk about general criteria used to evaluate computers, and the resources available to people with or without access to machines.

The temptation of most computer novices is to compare their own abilities to those of a computer—"The computer can alphabetize 10,000 words in less than 15 minutes, perfectly; even after a day and a half, I still have mistakes"—and then to conclude that the computer must be 216 times smarter. They couldn't be more wrong. The two adjectives that best describe a computer are "fast" and "dumb."

"Just hold on," the novice protests. "What about all those machines at the science museum that ask questions about your weight and food intake, and with the information garnered, evaluate your diet? What about my kid's SPEAK AND MATH that literally asks her questions and tells her when she's right or wrong?"

"You need to back up a little on that one," the computer buff replies. Inside the computer are microchips which code information. Someone has programmed the computer at the science museum, and someone else has programmed your child's SPEAK AND MATH toy, to respond in a particular way when certain information appears. Here is a sample of a simple program:

```
10 PRINT "What did you eat?"
20 INPUT N$
30 IF N$= "ice cream" then print "fatty"
40 IF N$= "carrots" then print "my slender sweetie"
50 END
```

What this says in language a computer recognizes is put the phrase "What did you eat?" on the screen. The answer you receive will be known as N$. If the answer (N$) is ice cream, then print back "fatty." If the answer is carrots, then print back "my slender sweetie." The numbers at the beginning of each statement simply order responses and commands.

Now here's an example of how dumb a computer is. If the question "What did you eat?" appears on the screen, and you write "icecream" or "iced cream," "peaches," or "spaghetti," the computer would send back a message saying that it couldn't deal with your response, or that the program was ended.

It isn't very difficult to get a computer to ask and answer questions, especially questions to which there are indisputable right and wrong answers. But to get a computer to replicate an apparently simple human action is much more difficult. If you want your child to pick up a carton of milk from the store, for instance, you say, "Here's $2, please pick up a carton of milk." To tell a computer/robot to pick up a carton of milk is incredibly complex. First, you would have to figure out what happens when a leg moves forward, how legs alternate, how and when the machine must turn when it comes to the end of a block, what to do to open a door (how to lift the hand and grab, how to turn a handle, how to pull toward you and move out of the door's way), how to find where the milk is located, how to access the money, how to wait for change, how to get out the exit door, which is different from the entrance door, and so on. Even if you were finally able to get these directions into computer language, the machine would be totally befuddled if it came to a door with an electronic eye instead of a doorknob. Moreover, if you want to tell a computer to do something as complex as "Go to the store" in discrete bits of information, the computer would need a huge memory.

Once a computer is programmed, however, it does whatever it is told to do very quickly; much more quickly than you or I ever could. In addition, the computer doesn't answer back with "Do I have to?" In fact, a computer programmed to go to the store will run back and forth between home and its destination until you tell it to stop, until it breaks, or until the conditions under which it is designed to run change.

Computer experts are generally interested, first, in how much memory an instrument has available, since they want to tell the machine to do lots of things, and second, in how quickly it processes information. All those words like bits and bytes, RAMS and ROMS, refer to the memory storage system or the speed at which information is processed.

Computer experts are also interested in the software, the information-filled disks one buys to run the machine. Telling a computer what to do, that is, using a series of commands, takes hours and hours of work and know-how. To save time and effort, computer users purchase other people's commands, and learn, for instance, what buttons the computer associates with "erase this word," "shoot down this alien," or "average these numbers." Software is generally computer-specific. The program disks you buy for an Apple cannot be used for an IBM. But the computer language you use with an Apple, that way of writing what the machine should do when X or Y happens, can be quickly adapted to most machines.

Don't be too easily impressed with what everybody else seems to know about computers. Because a person is able to push a canned program into a slot (or use a machine that's already been programmed, for example, a video game or math quiz) does not mean that he or she knows much, if anything, about computers. Jim Junior sitting at a keyboard, fingers flying and screens flashing, has learned infinitely more about DUNGEONS AND DRAGONS than he has about computers. The fellow in front of me at the computer supply store,

Consumer Checklist

√ Do not think of buying a computer in the way you think of buying a house. It is not a lifetime investment (in fact, it may be outdated fairly soon), and generally will not appreciate in value.
√ Do not assume that all computers work the same way. For instance, the video game programs from the computer mart will not produce graphics like those in a video arcade. Similarly, the word processing program you buy is probably not like the one in the office.
√ Do not assume that because a computer is capable of doing something you will be happy with the way it does it. For instance, a teacher put all his students' grades on the computer so that they could be averaged at any point in the term. He found, however, that the machine was slow in averaging grades and accepting new information; in short, the computer he owned did not do this job well. Other users have been equally disappointed with home bookkeeping programs and programs to keep track of cooking recipes. In fact, some computer experts go so far as to suggest that you begin by finding the software you wish to run and *then* look for the hardware that can run it.
√ Allow time to familiarize yourself with the machine's capabilities. For the first two weeks, for instance, it may take much longer to write a paper using the word processor than it does to do the same essay with pen and paper. A month later, you'll laugh at your former clumsiness.
√ Find out how the computer models differ, and then decide how important those modifications are for you. No two computer models are exactly alike.
√ In considering the cost of a computer, remember to figure in the cost of a screen, a printer, and software. Some machines come complete with these items, while others can be hooked up with devices already on hand (such as a T.V. screen or an electronic typewriter). In any case, check with a dealer before assuming that you're all set.
√ For a novice, look for a machine that is easy to use: one that comes with easy-to-understand documentation and that allows the user to correct mistakes with minimal effort.
√ In deciding whether or not to buy a computer, and about the choice of machine, (1) determine your needs and preferences and (2) understand what computers can and can't do for your child.

about to purchase double-sided, double-density disks (because they were on sale) for his TIMEX, knows even less. Lots of people who talk about the speed and power of their expensive machines may be correct in their assertions, but there is some question about whether the computers they buy are the best machines for them. It is senseless to purchase the computer equivalent of a Mercedes sportscar to drive back and forth on a one-mile bike trail.

Computer or consumer magazines, a good sales representative who carries more than one line, and a computer literacy instructor are your best bets for keeping up with available information. You should also take it as a given that you are not going to get the world's best buy on a machine—an excellent deal this week looks high to average six months later when prices drop. On the other hand, by waiting six months, you lose time that could be spent learning to use and enjoy the machine. My advice is this: Pick a price (when the X computer drops below $Y), or a time (when Janet goes into junior high), and then take the plunge.

PROGRAMMED LEARNING

In standard "educationese" (as opposed to computer talk), "programmed learning" refers to a system in which information is presented, then checked one step at a time through a prescribed set of questions and answers. For instance, in trying to teach the sound of short "a," the programmed text would present a picture of a bat and the words "bat" and "bit." The child circles one word or the other, checks for the correct answer, and moves on to the next problem. The system is premised on the notion that what is taught can be broken down into discrete steps with indisputably correct answers, and that immediate feedback is extremely helpful for the learner.

Although programmed texts are certainly available in workbook form, computers do a much better job of presenting this same kind of instruction: Children cannot "cheat" by looking at answers beforehand, and visual and verbal rewards for correctness are reliably forthcoming. In addition, sophisticated computer programs are able to skip ahead quickly for the child who, for example, gets five items of any kind correct, and to generate more and more items for a child who has not yet "caught on."

A large proportion (nearly 70% according to an analysis by the Educational Products Information Exchange) of the programs used by schools rely on this basic drill format. Similarly, most "educational programs" for home computers use this formula. Sometimes a clown will bow to a kid who's done math problems correctly, sometimes correct answers masquerade as alien starships to be exploded, but essentially the programs share much in common with the old flash card.

Although drill is one of the least impressive things a computer can do, many parents are pleased with programs of this sort. They want the computer to help their child do better in school, and the kind of reinforcement the machine provides often pays off in higher grades, especially in a school system which measures progress in similar ways. Moreover, computers can be as patient, consistent, and supportive as you program them to be. Computers also allow children to proceed at their own pace, and in this sense serve as individual tutors.

But there are important reasons why you may want to limit this use of computers. First, it is a questionable use of a computer's time; if there are children who are capable of doing something more interesting on the machine, why tie it up with a program that could be done equally well with flash cards? Second, although the computer is at first a motivational tool, children soon learn to hate a machine that insists on teaching them boring, rote material, in the same way that they hate teachers and tutors who insist on this approach. Children are thus "turned off" by a device which under other circumstances they would find fascinating. And, finally, children don't learn any more about the letter "E" just because a friendly dragon eats it, although this approach may lead them to expect their teachers to perform equally amusing tricks.

SOFTWARE REVIEWS

While textbooks and home learning programs are regularly reviewed, new software is, too often, merely celebrated. Computer experts generally comment on how the educational packages make use of the computer, or on the sophistication of the graphics, while little attention is paid to the content of the material. In evaluating reviews, note whether or not the expert comments on what the child is invited or expected to learn and how this approach compares to the same material in a printed format.

Software reviews from sources that specialize in noncomputer materials tend to be more sensitive to these concerns. The following well-regarded periodicals, among others, do a nice job with software: *School Library Journal, Booklist, Instructor Magazine,* and the several computer periodicals listed below under "Magazines." These publications should be available in most school libraries.

In addition, there are a few reviewing sources whose comments are particularly worthy of your attention. Several excellent books are especially satisfying for parents:

Only the Best: The Discriminating Software Guide for Pre-School–Grade 12 is available (for $19.95 including postage and handling) from

Education News Service
P.O. Box 1789
Charmichael, CA 95609

and *The Parent-Teacher's Microcomputing Sourcebook for Children* (Bowker, 1985) (available in most public libraries).

The National Education Association (N.E.A.) has designed a publication, *The Yellow Book,* which reviews 274 teacher-tested software programs, indexed and cross-referenced by grade (pre-school through high school) and subject. There is also a section on programs for the learning disabled. For further information, write:

N.E.A. Computers Service
4702 Montgomery Lane, Suite 1100
Bethesda, MD 20814

Institutions in search of reviews are generally well served by two other sources, *The Educational Software Selector* (T.E.S.S.) from the EPIE Institute (P.O. Box 839, Water Mill, NY 11976) and *Evaluator's Guide for Microcomputer-Based Instructional Packages* from MicroSOFT, a well-respected firm, located at 300 S.W. Sixth Avenue, Portland, OR 97204.

Instructor magazine also puts out a guide for educational institutions called *The Instructor Computer Directory for Schools* (Vogeler) for $20.

General comments on software, terse and sensible, are found in two sources which do not specialize in educational review. *How to Buy Software* by Alfred Glossbrenner (St. Martin's, 1984) is actually a compendium of tips: how to put a program through its paces before you buy, how to acquire free programs, what programs are good buys. *The Whole Earth Software Catalog,* Stewart Brand, editor (Doubleday, 1984) also offers specific reviews.

Directories of software for each machine are also available, usually in stores that sell your brand of computer. *Note:* Similar programs for different machines may vary radically in cost, so take this fact into consideration before purchasing one machine or another.

My own best advice on software comes from Robin Taylor, a specialist in evaluation. She rated the following ten companies as "most consistently producing programs of high quality."

Data Command
319 East Court
Kankakee, IL 60901

Edu-Tech
303 Lamartine Street
Jamaica Plain, MA 02130

Learning Ventures
P.O. Box 1860
Covina, CA 91722

South-Western Publishing
5101 Madison Road
Cincinnati, OH 45227

Strategic Simulations
883 Stierlin Road, Bldg A-200
Mountain View, CA 94043

EMC Publishing
300 York Avenue
St. Paul, MN 55101

The Learning Company
545 Middlefield Road,
Suite 170
Menlo Park, CA 94025

Odesta Corp
3186 Doolittle Drive
Northbrook, IL 60062

Spinnaker Software
1 Kendall Square
Cambridge, MA 02139

T.I.E.S. Project
1925 West County Road B2
St. Paul, MN 55113

I add to this list two firms whose work, particularly in elementary school science, is worthy of your attention:

Sunburst Communications
Room: BC 999
39 Washington Ave.
Pleasantville, NY 10570

Minnesota Educational Computing Corporation
3490 Lexington Ave. North
St. Paul, MN 55112

For a more comprehensive list of science-related computer programs, see *Computers in Science and Social Studies,* edited by David Ahl, from Creative Computing, 39 E. Hanover Avenue, Morris Plains, NJ 07950.

COMPUTER SIMULATIONS

Too many of the extant computer programs function as expensive flash cards, and too few are designed to take advantage of the computer's ability to model or simulate situations. There are, however, a few excellent programs that teach children concepts in this new and exciting way. In ODELL LAKE, a program for second- to sixth-graders, for instance, the user takes on the role of a fish in North American waters, seeking food while avoiding predators and danger. Either chance or poor reasoning can result in the protagonist's demise. This program comes on a disk called "Earth and Life Science" with four other science programs, all available through the well-respected Minnesota Educational Computing Corporation (3490 Lexington Ave. N., St. Paul, MN 55112) for about $50.

Other simulation programs have become popular in the teaching of physics. Because physics focuses on how objects move, theoretical schemes can be demonstrated by computer that would be well-nigh impossible to show with hands-on experiments. Although these programs work extremely well in schools, they are not particularly useful for a parent who wishes to purchase software that children will use over and over again.

In evaluating simulation programs, be concerned that the simulated activity is portrayed correctly. For this purpose, I tend to trust reviews by subject specialists and educators rather than computer people who are impressed with the graphics. Also, be suspicious of simulations that teach, either overtly or subconsciously, that life is merely a game. It is finally dangerous and unproductive to see all humans as absolute friends or enemies, enemies who have no rights, no feelings, and only a limited series of "moves."

Note: Simulation programs can usually be run only on machines with large memories.

TEACHING LOGIC

The best computer programs I've reviewed are designed to teach logic, or what is sometimes called "the scientific method."

INCREDIBLE LABORATORY for instance, invites users to figure out which "strange chemicals" cause which monster features. This award-winning program, designed for children in grades 3 and above, begins with five elements to sort out; at the expert level one works with fifteen unknowns. The graphics in the game are excellent, and the program is challenging enough to require serious note-taking by answer-hungry users.

Sunburst (Room: BC 999, 39 Washington Ave., Pleasantville, NY 10570), creator of **INCREDIBLE LABORATORY**, also produces at least two other excellent logic programs, **THE FACTORY** and **THE POND**, which work on the Apple, Atari, Commodore, and IBM (all run about $55).

COMPUTER LANGUAGES: LOGO

"Programming" a computer is another word for telling the machine what to do. For children interested in learning a computer language there are several choices. The most commonly taught programming language is called BASIC; as its name suggests, this language gives users a not-too-difficult way to make a computer do what they want it to do. Variations of BASIC are available for virtually all home computers. If one is displeased with the enclosed directions, instruction manuals can be purchased in bookstores or borrowed from libraries. Some of these guides are computer-specific, while others attempt to give the user an overview of procedures, with particular advice on how to adapt what they have learned to their own machine.

To program in BASIC one must be able to read and type, at least at an elementary level. There is, however, another computer language, LOGO, which allows even young children to program. LOGO, specifically the part referred to as "turtle graphics," invites users to draw by giving a turtle mathematically sophisticated directions. A "TURTLE TOT" robot is also available: the same mathematical direction-giving results, in this case, in the robot moving. LOGO is accessible to small children, but do not assume that it is simple. The language teaches logic and allows a user to draw an amazing variety of forms. Although LOGO is an interesting program language for adults to learn, most grown-ups head toward BASIC or one of several of the other programming languages which are commonly used in adult work.

I like LOGO. It teaches children that they are in charge of the machine, while other programs, the rote learning ones in particular, often have children believing that the computer is smarter than they are. Moreover, LOGO is available to non-readers who are ready for a computer experience but unable to handle BASIC. In addition, LOGO is clearly imbued with an aesthetic dimension; as children draw with their turtle they are at once amazed at what they have done and ready to make decisions about what to do next. Both children who are excited about learning a computer language and those who want to create a graphic image are attracted by LOGO.

Because LOGO is basically a graphics program which requires color and a significant memory, the software is not available for all machines. Although the original Terrapin LOGO is generally considered the best version, Krell LOGO is also good. See if one of these is available for your machine.

The *National LOGO Exchange,* a newsletter concerned with practical LOGO teaching techniques published September through May, has been helpful for parents as well as teachers. Not ony does this periodical include articles, reviews, and evaluations, but it also serves as a LOGO network and research directory. For more information write: The National LOGO Exchange, P.O. Box 5341, Charlottesville, VA 22905.

I have actively sought out criticisms of LOGO, mostly because as its adherents tell the story, it seems almost too good to be true. There are a few people who claim that because LOGO is so much fun and the satisfactions are so immediate and great, children learn to expect that all computer programming will yield these kinds of immediate rewards. When they don't, former LOGO students are more likely to quit. LOGO teachers, on the other hand, claim that children who have begun with BASIC have a more mechanical approach and have a harder time learning LOGO than children who are starting fresh. Personally, I am astonished and delighted to see five- and six-year-old children type in complex directions for their turtle computer.

Although there are other computer languages available—FORTRAN, COBOL, and PASCAL are three frequently used in engineering, business, and science—these are generally not taught to pre-high-school students.

THE COMPUTER AS TOOL

Computers are versatile tools. The writer who uses a word processing program and the scientist who converts data from an audible to a visual signal are both using programs that give them a flexibility and scope they would not have without these instruments. But while a word processing program is adaptable to most written compositions, scientific programs tend to be either extremely complex or specifically adapted to a particular machine or problem.

Learning to use a computer as a tool takes time. The third-grader who is able to put all his baseball cards into a data base and then call up players by year, team, and so on is certainly not "doing science," and yet the skills he has learned in this activity may serve him well later on. Similarly, a child who can design and correct a program so that the building she draws on her screen "looks right" is also learning important computer skills which may serve her well in science.

In other words, it is unlikely that a child not yet in high school will be able to calculate the orbits of planets or construct the circuitry needed to make a robot's arm move up and down. On the other hand, the high-schoolers who accomplish these feats for science fairs have been "playing" with computers for years—familiarizing themselves with the keyboard, programming techniques, and, finally, coming to understand the power of the machine.

In some software packages a number of functions are combined. For instance, SKY TRAVEL, an excellent program from Commodore for children age twelve and older, uses sophisticated astronomical calculations to play a game which involves problem-solving. During play, important factual information is reinforced. I had trouble with this game, not because the computer functions were difficult, but because I didn't know enough astronomy.

HANDS-ONS

Even if you are not ready to invest in a computer, there are a number of toys on the market designed for kids who have been bitten by the bug and are not prepared to take the cure. The worst of these attract children by having pictures of computers on the box covers while the contents are only vaguely related to hands-on use—flashcards with drawings of computer parts to identify, board games based loosely on circuitry patterns, and computer-related words to identify.

There are, however, several excellent toys which invite a child to use many of the same thought processes that one uses on a computer. The simplest of these are logic toys.

PENTE (Box 1546, Stillwater, OK 74076), an ancient board game, uses a reasoning process that provides the kind of mental exercise a future hacker finds stimulating. **THE "L" GAME** (Jabo, Inc., 1126 Clifton Rd., N.E., Atlanta, GA 30307), another seemingly simple logic toy invented by psychologist Edward de Bono, provides a similar kind of practice. Both of these are available in many museum and gift shops.

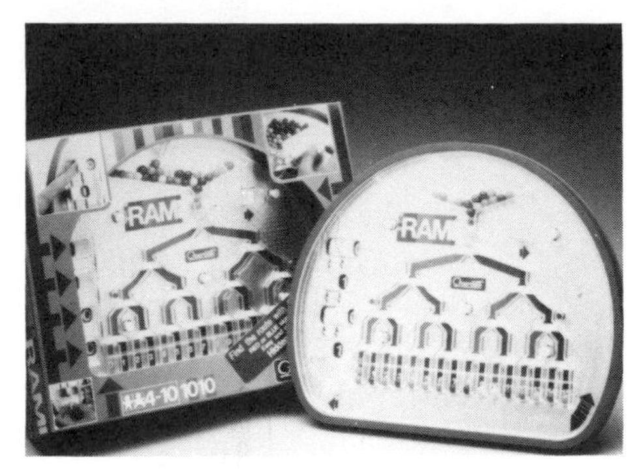

Fig. 91

Fig. 92

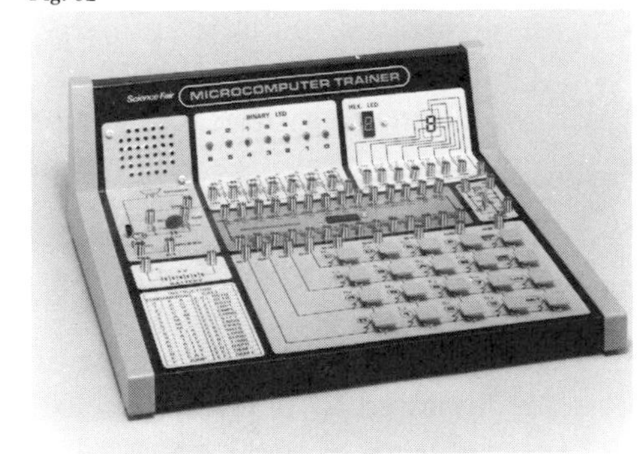

Fig. 93

Fig. 91. **RAMI,** a binary math game for children ages six to eleven, gives young people practice in the math that computers use in programming. This game comes with a cover so that after a child has mastered the basic binary code, he/she can try the system using only numbers. Available in better toy stores and catalog houses for approximately $19, this one is a perfect toy for kids who like mazes and delight in abstract reasoning.

Fig. 92. A **DIGITAL "COMPUTER" KIT,** available from Radio Shack for $27, is designed to help children learn how machines are programmed. This simplified computer, which works on a binary code, can actually be wired to play a number of riddle-type games, and function as a vocabulary building device, a question-and-answer board, and more.

Fig. 93. The **MICROCOMPUTER TRAINER** from Radio Shack, which sells for $30, is designed to explain binary and hexadecimal math, and to teach beginning assembly language through a series of games. While for most purposes the languages such as BASIC and PASCAL, discussed above, are perfectly adequate, machine language takes the user right to the circuits and microchips. When mastered, machine language makes programming faster and gives users access to the elements that make all programming possible.

Fig. 94. A **DIGITAL LOGIC TRAINER/COMPUTER LOGIC LAB,** imported as an N.S.I. exclusive from the Thomas Salter Co. of Scotland, provides a series of experiments to teach the workings of a computer. This kit will not necessarily make a child more skilled at playing computer games or mastering "drill and practice" programs. Rather, it is designed to help users understand how, for instance, a buffer stores information, or the precise function of an LED. Approximately $25 from toy stores and hobby shops; recommended for children age ten and older.

Fig. 95. The **MEMOCON CRAWLER,** a member of the **MOVIT** robots series from O.W.I., is characteristic of a new breed of toys which can be interfaced with computers. This particular version, which retails for approximately $75, allows the assembler to instruct the 4-bit microprocessor brain of the robot to follow up to 256 steps. For further information, write: O.W.I. Inc., 1160 Mahalo Pl., Compton, CA 90220.

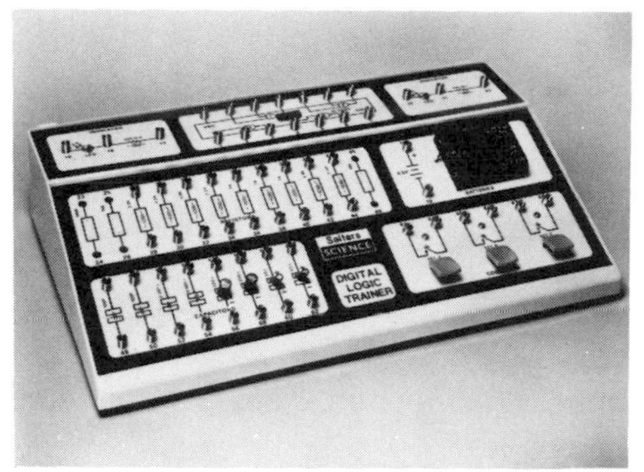

Fig. 94

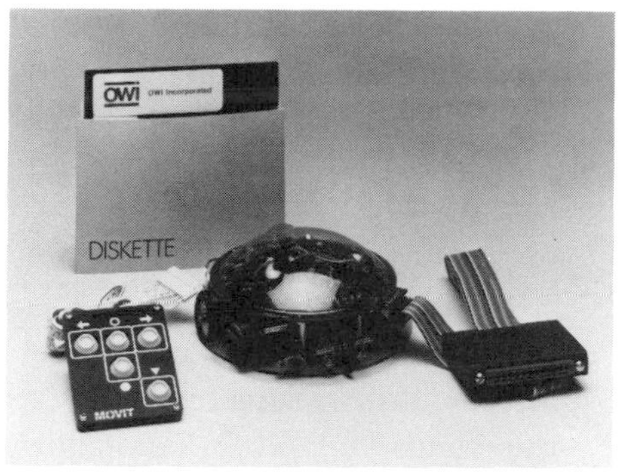

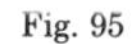
Fig. 95

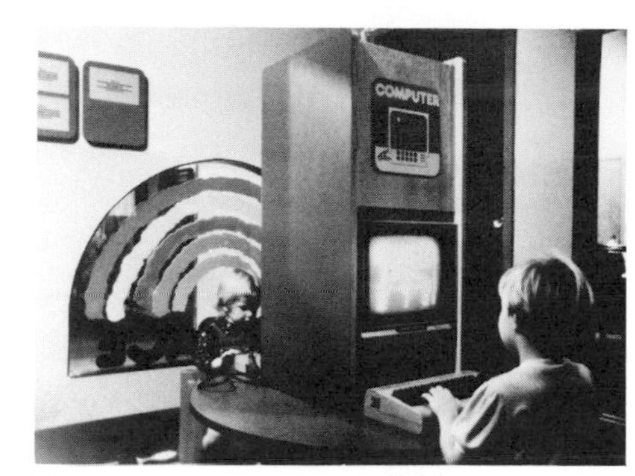

Fig. 96

Fig. 96. Computers are increasingly and generously sprinkled in spots children visit: schools, libraries, museums. The children shown here are working with games at the Center of Science and Industry, Columbus, OH.

FINDING A COMPUTER: ALTERNATIVE STRATEGIES

Despite the challenges and pleasures provided by computer-related toys, for children who want to learn about computers there is no substitute for extensive training and practice. Computer teachers regularly report that those students who have access to computers at home, and have the time to practice and play and invent, are at a tremendous advantage in class. If you can't afford a computer yourself and want your child to have time to practice on a good machine, there are several strategies worthy of your attention.

- Find out how many computers are available in your child's school. Although these may be used heavily during peak hours, suggest that during lunch, after school, or on weekends, an aide might be hired to let motivated and trustworthy children have access to the machines.
- Check on the possibility of getting your library to make a computer available to patrons.
- Investigate computer camps. They have become a popular alternative to in-school instruction.

- Find out if local industry is willing to set up computer tutorial sessions on weekends or evenings.
- Call a local college and ask if courses are offered for kids through the education department or computer department.

IN PRINT

Ault, Roz, *Basic Programming for Kids* (Houghton Mifflin, 1983).
Specific instructions for seven of the most popular home computers make this hands-on guide a hands-down winner.

Bearden, Donna, *1, 2, 3, My Computer and Me* (Reston, 1983).
LOGO explained to children in language simple enough for their parents.

Bly, Robert W., *Ronald's Dumb Computer* (Dell, 1983).
This fictional work provides a basic introduction to the youngest would-be hackers on programming, flow charting, and logic.

Bolognese, Don, and Robert Thorton, *Drawing and Painting with the Computer* (Watts, 1983).
The graphics capabilities of home computers are explored in this well-illustrated volume.

Freedman, Alan, *The Computer Glossary* (Prentice-Hall, 1983).
Both a comprehensive reference book and a guide to computer literacy, for advanced readers.

Hardy, Jack B., *Adventures with the Atari* (Prentice-Hall, 1984).
A creative, well-organized guide to writing interactive and graphic adventure game programs. Athough designed especially for the Atari, the procedures apply to all microcomputers.

Lipson, Shelly, *It's BASIC* and *More BASIC,* (Holt, Rinehart & Winston, 1983 and 1984).
Each book offers a straightforward approach to the teaching of BASIC to children in grade 3 and above.

Markel, Sandra, *Kids' Computer Capers: Investigations for Beginners* (Lothrop, Lee & Shepard, 1983).

Mysteries to solve, surprising facts, and challenging puzzles mark this clever introduction to computer literacy.

Nozaki, Akihiro, *Anno's Hat Tricks* (Philomel, 1985).

Binary logic and mathematical deduction are used to figure out the color of the hat that each of the characters is to wear. Genuine mind-stretching activities for the youngest of mathematicians.

Ramella, Richard, *Computer Carnival: Sixty Programs for Starters* (Wayne Green, 1982).

A wide range of short computer programs (both educational and fun) for the TRS-80; they can be modified to other machines.

Simon, Seymour, *How to Talk to your Computer* (Crowell, 1985) and *Meet the Computer* (Crowell, 1985).

These companion volumes for computer-interested children in grades K–3 are clear, attractive, and thorough. Sequential language, BASIC, LOGO—it's all here.

Targ, Joan, *Ready, Run, Fun: IBM PC Version* (Prentice-Hall, 1985).

Targ assumes no prior knowledge of the machine. She competently and enthusiastically leads readers through a series of relevant programs that teach important concepts—drawing simple pictures, writing thank-you notes, and games. An excellent resource for IBM owners.

Walter, Russ, *The Secret Guide to Computers, Vol. I* (Birkhauser Boston, Inc., 1984).

For teenage beginners, there is nothing better on the market—challenging, practical, and altogether fun.

For Parents

Galanter, Eugene, *Kids and Computers: The Parents Microcomputer Handbook* (Perigee, 1983).

Sensible advice from a professor of psychology who runs a computer school in New York City.

Papert, Seymour, *Mind-Storms: Children, Computers and Powerful Ideas* (Basic, 1982).

A LOGO classic, Papert's provocative book describes the theory and practice of LOGO as studied by members of the Artificial Intelligence Lab at M.I.T.

Williams, Frederick and Victoria, *Growing Up with Computers: A Parent's Survival Guide* (Morrow, 1983).
Anxious readers are given not only good practical advice, but also an important attitude toward the machine and the children who use it.

Computer Magazines

Because of the rapidly changing parameters of the industry, the magazine format is particularly useful to the computer buff. Although there are numerous periodicals designed for sophisticated computer users, these six are especially for kids and their families.

Digit, 6/year, $9 (P.O. Box 29996, San Francisco, CA 94129).
The perfect computer magazine for a kid already excited by ideas and ready for more challenges.

Enter, 10/year, $13 (One Disk Drive, P.O. Box 2686, Boulder, CO 80322).
This periodical from Children's Television Workshop is as upbeat as the *Electric Company.* Computer games and other trendy topics are often highlighted, and computer applications are regularly covered. With only introductory work on a school computer, one could enjoy this magazine.

Family Computing, 12/year, $18 (730 Broadway, NewYork, NY 10003).
The only computer magazine designed for the entire family. Software reviews found here are really for home (as opposed to schools), and the monthly puzzles invite adults and kids to share in problem solving.

Microkids, 6/year, $15 (P.O. Box 992, Farmingdale, NY 11737).
Microkids is to computers as *Science Digest* is to science. This light journal is more about computer culture than computers and their capabilities.

Microzine, 5/year, $149 (P.O. Box 947, Education Plaza, Hicksville, NY 11802).
This computer magazine on disk wins high praise from experts and kids alike. Regular features include a story where the reader helps determine the plot, an interactive interview (yes—your child participates), and programming instructions for games and data storage.

20 Building and Engineering, or, Look What I've Made!

So many little pieces. So many brands. So many colors and locking systems and add-ons. For parents who stand immobilized before the construction section of a mega-toy store, let me offer a friendly bit of advice: No young person needs, or should have, all these items. They would be as confused as you are standing there! Clarify your goals. You want materials that allow children the various pleasures of building: the satisfaction of making something of their own; the living proof that scientific principles work; and the smile that comes from standing back and gawking at an impressive object they, themselves, have made.

HANDS-ON

Construction Toys

Wooden Blocks

Promoting a good set of wooden blocks is like lauding apple pie and motherhood—there is little disagreement about their value. This is not a sarcastic comment; if you have the money, buy a set. The perfect set of wooden blocks includes a variety of shapes that are proportionally designed—that is, two squares are the same size as one rectangle—and are fairly heavy (they balance better if they are heavy). In short, the heavy oak blocks are best for building. Complement these with brightly colored blocks, play people and animals—all of which work well as decoration.

The principles learned in block play are fairly simple. Blocks which are balanced, or sit end to end, provide a stable foundation; unbalanced and poorly supported structures tumble. There are no fixed supports available in a block set; in engineering terms, the blocks cannot be cantilevered by locking, tying, or gluing them in place. Construction toy developers, beginning where wooden blocks leave off, have produced a number of construction sets which do what wooden blocks cannot.

Cantilevered Blocks

There are a number of building systems that use a locking arrangement. Each of the following offers children certain advantages.

Bristle blocks are colorful plastic blocks covered with protruding nodules which literally stick to any other **bristle block**. They come in a variety of shapes, are relatively inexpensive, and work well for even a young child who has trouble mastering the interlocking devices used in other systems.

Duplos, the large plastic bricks and oversized interlocking nodules, help small children make barns, trains, and houses. Many people do not realize that **duplos** are made by the **lego** people and can be locked into the smaller **legos**, making the transition from gross to fine building easier. **duplos** also come with people, animals, vehicles, and buildings designed to interlock with these same basic bricks.

Fig. 97

Fig. 97. **Construx,** an excellent building set from Fisher-Price, uses members that resemble gray steel beams and joins them with plastic connectors that form right angles. Several other hinged and pivoting pieces and a pulley are included in the larger sets. Because the beam lengths have been thoughtfully designed, children soon learn that additional members can be added for support and reinforcement. Finished products are indeed stable and very attractive. A rewarding system for five- to eight-years olds.

Erector sets, originally designed by A. C. Gilbert in 1913, are what most of us recall as an advanced construction toy. The system uses a series of metal pieces which are bolted together to produce free-form angles. Railroad tressles, towers, and bridges look particularly realistic. Ideal, the new owner, now includes motorized units, remote control devices, and pre-fab chassis to make construction easier, less free-form, and more in keeping with the space age. Because the nuts and bolts are small, and many of them are needed to construct anything, the set is time-consuming and difficult for children under ten to put together.

Fig. 98

Fig. 98. **FISCHERTECHNIK,** a truly comprehensive building system from West Germany, is designed to teach scientific principles and applications. These kits provide modular components that can be combined to form both static structures and functioning model machines. The starter sets, designed for children in kindergarten and above, include gears, axles, structural members, and connectors, and sell for about $40. The most advanced sets—the electro-mechanical kit ($75), the pneumatics kit ($75), the compressor ($53), the statics set ($43), the electronics kit ($95)—can be used by competent sixth-graders, but are often purchased by universities and industrial firms working on design problems. A spectacular and very expensive system. As pictured, the FISCHERTECHNIK components are being assembled on the **TABLETOP TECHNOLOGY SYSTEM** designed especially for them by Creative Learning Systems, Inc. (9889 Hibert St., Suite E, San Diego, CA 92131).

Fig. 99

LEGOS, the small interlocking blocks which culminate in the **EXPERT BUILDER** series, are sold through toy stores nationwide. **LEGO** makes available hinges, pivoting mechanisms, and a variety of motorized parts. Because of the small size of the blocks, one can achieve a detail not possible in the sets with larger members.

Fig. 99. In addition, **LEGO** makes a series of school sets available to the public through either their Connecticut office (555 Taylor Rd., Enfield, CT 06082) or Transtech. Designed to teach specific physical principles with levers, belt drives, gears, and so on, the school set gets excellent reviews from both children and engineers. Each model takes about 20 minutes to build. Directions explain the theory involved in the construction and real-life applications of the principles involved, suggest experiments, and provide definitions of terms and mathematical formulae when pertinent. The 179 LEGOS included in the school sets are the same blocks used in store-bought LEGO sets, so they can be added to pre-existing projects. These kits are not blister packed, they are not commercially advertised, and thus children do not know to request them at gift-giving times. Recommended age: eight to fifteen. Price: about $35; teacher's manual, $4.

LOC BLOCS use the same basic system as LEGO for cantilevering members, but the sets sell for considerably less. LOC BLOCS cannot be used interchangeably with LEGOS, and are best viewed as a competitive building system. Their EXPERT BUCKET, with 740 members, features wheels, axles, roofing blocks, mini-people, doors, and more, but not as many devices as are included in the LEGO line.

Fig. 100

Fig. 100. Tyco SUPER BLOCKS do work with LEGO blocks, but cost considerably less. Tyco also makes a PRE-SCHOOL BLOCK which is interchangeable with the DUPLO block. For someone who has LEGOS and wants more blocks to build with, this is the way to go. New sets of SUPERBLOCKS, which can later be combined with the high-tech LEGO accessories, range in price from $5 to $30, less applicable consumer rebates.

LINCOLN LOGS, the notched logs with which one builds model cabins, are most successful with a child dedicated to playing cowboys. Although the cabins are fun to construct, consumers should realize, first, that it takes a lot of logs to become a creative cabin builder, and second, that the structures are not, by contemporary standards, stable.

Fig. 101

Fig. 101. NUTS AND BOLTS, and several other attractive and interesting toys which use the same screwing mechanism to connect wooden pieces, may seem tedious to a child accustomed to simply snapping members in place.

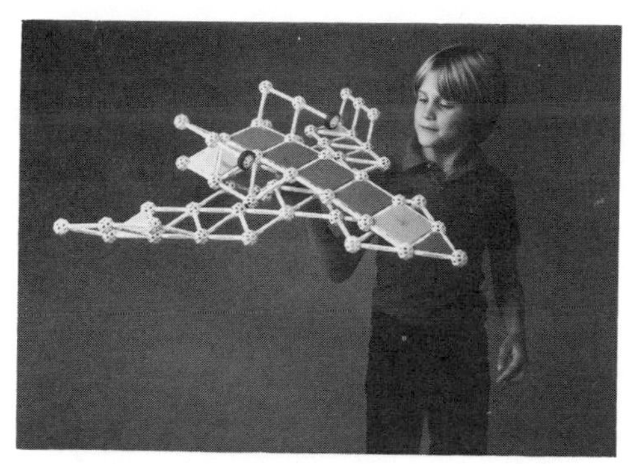

Fig. 102

Fig. 103

Fig. 102. **RAMAGON** is the only construction set that invites children to form 45°, 90°, 135°, and 180° angles. A 26-sided ball with one hole on each side is used to hold rods and clips as kids make pyramids, tetrahedrons, parallelepipeds, and other shapes not possible in building sytems based on the 90-degree angle. **RAMAGON** also sells panels with raised nodules designed to accommodate **LEGO** blocks. This system is stable, attractive, and satisfying.

TINKER TOYS are the only building system which includes a wheel and a simple push rod. Although the windmills are neat, our child reviewers had trouble getting certain sticks to fit and keeping constructions, once built, intact.

Fig. 103. The **GIANT TINKERTOY BUILDING SET** is, without question, the largest construction toys on the market. The 53 piece set, designed for children ages 3½ and up, retails for $46.

In choosing a set of interlocking blocks for your children, consider their interests, abilities, and your long-range building set plans. If, for instance, you eventually want a set with motorized parts and other high-tech extras, begin purchasing nonmotorized pieces from that company when your child is six or seven. If, on the other hand, you believe that your child will outgrow blocks before reaching the age when this large investment in small pieces pays off, go with a less expensive version. In many ways, buying toys is like planning an investment strategy: you can only guess whether interest rates will climb or drop in the months ahead.

Motors and Gears

Children who like building enjoy having at least one toy that invites them to learn about movement.

The simplest of these kits is called **LOTS O GEARS**, and is designed for children ages four to seven. Colorful gears of different sizes interlock with Velcro anywhere on the board. Children turn one gear and watch others move accordingly. This toy, which sold for about $17.50 at toy stores and catalog houses, is no longer being produced, so if you see one and you're looking for an aesthetically pleasing, hands-on experience with gear ratios, grab it.

Fig. 104

Fig. 105

Fig. 104. **CAPSELA,** another highly recommended toy, takes an entirely novel approach to the notion of construction. Instead of trying to replicate objects that children can already identify—steam rollers, ferris wheels, and space ships—CAPSELA begins with the notion of function. Within clear plastic bubbles, Japanese engineers have encased motors, worm gears, and gear reduction devices which invite young people (and adults) to build their own land and water vehicles.

Fig. 105. **CAPSELA'S** programmable toys, the 2000 and 5000, which snap onto space-age vehicles, allow children to type in seventeen different commands. These models, which come with sound, light, two motors, and chain drive can be made to move forward, backward, turn, pivot, and travel a hundred feet.

CAPSELA is an impressive toy. The plastic bubbles are at once strange and beautiful. Moreover, because CAPSELA is not a model kit, it encourages a kind of creativity not promoted by the step-by-step directions available with other building toys. And, finally, I like to watch kids work with this one; not only do users describe their productions with enthusiasm, but they also learn to use notation, and to vary interchangeable working parts in scientifically and technologically interesting ways. The largest CAPSELA motorized constuction system retails for about $80, although starter kits are available for considerably less. CAPSELA also produces little people that fit in the capsules for children who create astronaut-filled stories with their vehicular creations.

For the truly advanced builder, I recommend three difficult and interesting sets.

A **VISIBLE V-8 MODEL ENGINE,** available through better toy stores, includes over 350 parts, 100 of them moving. Once completed, you see how spark plugs fire in order, pistons drive connecting rods and turn the crankshaft, and valves and belts are synchronized for efficiency. This set sells for $30. The VISIBLE V-8 engine, a ¼ scale, transparent plastic model, actually works.

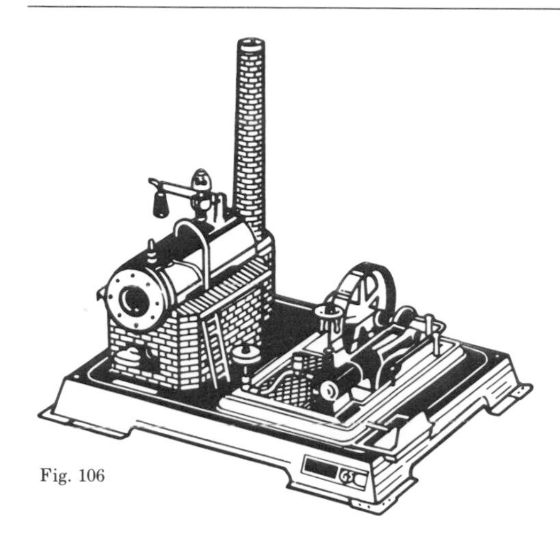

Fig. 106

Fig. 106. Wilesco, a West German company, makes an extremely precise replica of a steam engine that allows builders to see what happens when "fire and water" are brought together to produce mechanical energy. The Wilesco engine can be used to drive a miniature drilling machine, a saw, a locomotive, a steam roller, or anything else you wish to power. Because the machine burns solid fuel pellets, and because it does produce real steam, adult supervision is essential. This Wilesco steam engine sells for about $50 in hobby shops.

The most difficult of the construction toys is produced by Mantua, a series of HO-gauge working locomotives. These require specialized tools—a small block of steel which serves as an anvil, a needle file, a center punch, a light hammer, and a place out of harm's way to build the thing. The MANTUA STEAM LOCOMOTIVES sell for $30 to $100 from Mantua Metal Products Co. Inc., Grandview Ave., Woodbury Heights, NJ 08097, or through hobby shops.

Geometric Forms

The Franklin Institute Science Museum of Philadelphia produces several engineering kits. Although the included materials are readily available—coffee stirrers, pipe cleaners, tongue depressors, cardboard—the consumer is actually purchasing a design and a fine explanation from the Museum Education Office. The following engineering kits are now available:

Arch Bridge, $2.98
Geodesic Dome, $2.98
Space Frame, $2.98
Tents, $6.95
Truss Bridge, $6.95

Children should realize beforehand that the objects they build will not look fancy or "finished." They should also have some plan for the completed object—the resultant structures are fairly large and not particularly attractive.

To work with these kits independently, children ages eight to eleven need good eye-hand coordination and a real interest in the subject. For purchasers of these kits, Tip #1: The rubber washers used in the Truss Bridge must be pushed on dome-side first. Tip #2: If your child is bothered by the flimsiness of the pipe-cleaner construction, try tying or gluing the members in place.

Sets can be purchased by writing directly to The Franklin Institute Science Museum, 20th and The Parkway, Philadelphia, PA 19103; add 50 cents shipping and handling to all prices. Several other museum shops are also carrying The Franklin Institute line.

Another educationally oriented organization, The Buckminster Fuller Institute, also sells items of interest to the young engineering enthusiast. Although dome fans should write directly to the Institute for a catalog (1743 S. La Cienega Blvd., Los Angeles, CA 90035), several items produced there are worthy of particular note. The **GEODESIC DOME KIT**, which sells for $14.95, makes two 3′ diameter geodesic domes or one full sphere. Kids not only come to realize how strong the dome shape actually is, but can also use the structure as a playhouse. The kit is easily assembled, although young children need adult help.

Other dome kits are available in science and hobby stores nationwide. In evaluating them, look at the way the members connect; ideally these sets should be easy to build *and* solid.

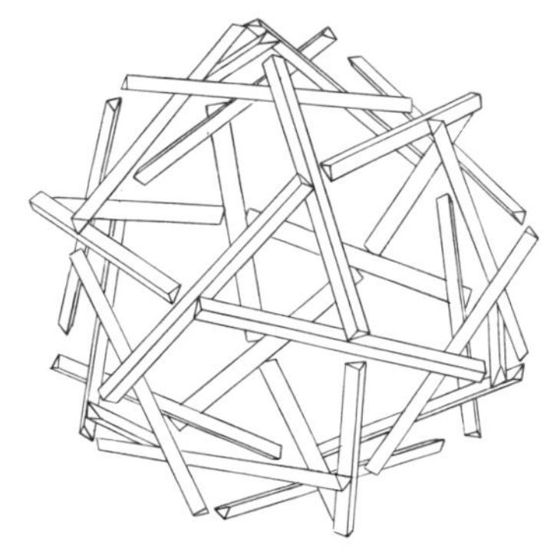

Fig. 107

Fig. 108

Fig. 107. **SPACE BEAMS,** created by engineer Edward Gorczyca, are designed to help young people discover "statics." Although the directions are a bit complex, and considerable dexterity is needed to complete the model, the finished product made from either clear acrylic "straws" or heavyweight acrylic rods dangles attractively in space in a sunny window. **SPACE BEAMS** are sold in several museum shops, or can be ordered directly from Go-Co, 2905 Hemlock Dr., Allison Park, PA 15101. The simplest set sells for $4.85, and a more complex version, which uses thirty triangular rods, can be had for $17.85. (Prices already include postage and handling.)

Fig. 108. A relatively inexpensive series of three-dimensional kits, sold in museum shops and toy stores, is produced by Ikoso (28667 Spencer Creek Road, Eugene, OR 97405). The kits include thin, round, wooden struts, clear plastic connectors, and a push pin to make holes in the connectors. The kits are recommended for older children—it is difficult to make the requisite holes and cut the struts to the correct size. The sphere kit took two full evenings, and one blister, to complete. For the right person, however, these sets are packaged nicely and make good stocking stuffers or non-elaborate birthday presents. The prices are approximate.

Discovery #1, $1.50	Black Mobile, $4.50
Stars and Crystal, $2.00	Platonic Mobile, $4.75
Small Sphere, $3.00	Great Star, $2.75
Large Sphere, $4.50	Superkit, $5.00

Building with Wood

There are a number of self-contained tool/building kits on the market, wooden cabinets neatly packed full of hammers, chisels, and saws. To make a long story short, the miniature tools are cute but ineffective. It is difficult to get equipment back into the box, and the tools included are not all the ones children need for the projects they wish to build.

People adept at encouraging children's carpentry offer the following suggestions.

- When a child is born, buy him/her an empty tool box. For birthdays or holidays, add one tool at a time. An aunt and uncle might well prefer to purchase a good chisel instead of another stuffed animal.
- Use professional tools, even for young children. There is more danger in trying to get an inadequate, dull saw to work than there is in using a sharp instrument correctly.
- Make sure that the child has a place to work that is out of the way, can be made messy, and can be cleaned up again. Some parents consider building an outside project, while others have set up a spot in the family work area for another set of hands.
- If you have enough bench space, build your child a movable platform so that he/she can work comfortably at adult height. Workbenches should be about 24″ to 30″ high, depending on the height of the child.
- Help the child organize a workplace so that tools can be put back in an

Richard Starr, author of *Woodworking with Kids* (Tauton Press, Box 355, Newton, CT 06470, 1982), summarized his advice on tools in an article, "Teaching Your Children the Joys of Woodworking," in the December 1983 issue of *Handyman.*

Here is what he considers a basic complement of tools:

Hammers An 8- or 10-oz. hammer or a small ball peen or tack hammer. These effectively drive 1½″ long finishing nails into wood ¾″ (or less) thick.

Saws (a) A crosscut saw with fairly fine teeth, 10 teeth to the inch or finer, (b) an 18-in. carpenter's saw, (c) a good coping saw with a stiff frame for high blade tension (mount the blade with the teeth pointing toward the handle), and (d) a dovetail saw for small work.

Drills A quality, well-oiled hand drill is important; cheap models stick and jam. For holes up to ¼″, use a hand-powered gear drill with standard twist bits. For holes larger than ¼″, use a brace and bit.

Measuring tools (a) A combination square or try square, for laying out right angles, (b) an 18″ or 24″ ruler, and (c) a compass.

Shaping tools Files, rasps, and other surface shaping tools help round and smooth edges. Planes, which are somewhat difficult to use, help a child understand much about the nature of wood—begin with a no. 3 (8½″) bench plane. Block planes are useful, but easily damaged. A tiny palm plane works well for beveling edges. It is important to keep all of these tools sharp.

Vises and clamps Children soon learn that a vise is like an extra set of hands; mount the edge of the vise flush to the corner of the workbench, so your child can saw close to the vise. In general, C-clamps or quick-set clamps work better for young children than large heavy clamps.

accessible spot and found again when needed. Outlining items on a pegboard, for instance, is an effective strategy.

- Do not let a young child work with power tools. There are plenty of great projects to be made without them.
- Do not let a child work in a shop unless an adult is within earshot or actually supervising, depending on age and ability.

Cardboard Carpentry

For those who hear "cardboard carpentry" and conjure up an image of supermarket boxes ripped and taped together, think again. Cardboard construction is a fascinating alternative to wood carpentry: cardboard is cheaper than wood, and in many instances, as stable; the cuts are easier to make if one has the right equipment; and the materials themselves are easier to store and lighter to move. And because this system allows even young children to produce usable, attractive products quickly, they can begin thinking about the larger questions of function and design, instead of working furiously for 15 minutes to get a nail in "right."

Cardboard projects are made from tri-wall corrugated cardboard, available in sheets from Learning Things, Inc., 68A Broadway, P.O. Box 436, Arlington, MA 02174. This company also sells a set of unique tools—sharp knife blades mounted so that V grooves, slots, circles, and other cardboard shapes can be made easily. Most important are the books of project directions they sell, *Cardboard Carpentry* and *The Further Adventures of Cardboard Carpentry* by George Cope and Phylis Morrison (1973). Although the cardboard is much cheaper than wood, there is a significant initial expenditure for tools required. I recommend cardboard construction for schools, Scouts, or for communities who want to buy tools and share them.

Figs. 109–110. Cardboard carpentry allows elementary-school children to construct workable, strong bookshelves, chairs, sand tables, bed frames,

Fig. 109

Fig. 110

wagons, playhouses, and more. A beginning kit that includes tools, tri-wall, wooden elements, and directions retails for about $200 from Learning Things. Additional tools and guidebooks are also available.

Drafting Sets

Drafting is a regular feature of science education in the Soviet Union, but here drafting is more often likened to art. Scientists and engineers queried on this point unanimously agreed that they wish they had taken drafting earlier on. Moreover, they saw it as an extremely useful and satisfying skill.

There is only one drafting kit, made by Crayola, designed for young people. Fortunately, it is excellent. The equipment is good, the directions are clear, and the finished products are astonishing. Both children and adults who work with these sets come away feeling that they have learned something they hadn't even had an inkling of before. My only complaint is the sexist packaging.

Fig. 111

Fig. 111. Top honors for the CRAYOLA DESIGNER KIT, intended for children age seven and up. A well-designed, educationally sound, and satisfying toy which retails for about $19.

IN PRINT

For All Ages

Keller, Charles, *The Best of Rube Goldberg* (Prentice-Hall, 1979).
Rube Goldberg, a cartoonist, created zany and complicated machines in line drawings. They remain clever, funny, and always intriguing.

Macaulay, David, *Castle, City, Underground,* and *Unbuilding* (Houghton Mifflin, 1973, 1974, 1976, and 1980).
Macaulay, an architect by training, produces spectacularly imaginative, informative, and beautiful books. Highly recommended.

For Young Children

Bushey, Jerry, *Building a Fire Truck* (Carolrhoda, 1981).
A complete and instructive description of how fire trucks are made, and a real winner with fire truck fans.

Gibbons, Gail, *Tool Book* (Holiday House, 1982).
Common tools and their uses are boldly, artistically, and clearly presented here.

Mitgutsch, Ali, *From Clay to Bricks, From Ore to Spoon,* and *From Sand to Glass* (all Carolrhoda, 1981).
These books about basic processes are accurate, easy to follow, and interesting for child and adult readers.

Thomas, Art, *Merry-Go-Rounds* (Carolrhoda, 1981).
The merry-go-round began in Arabia more than 900 years ago. To find out more, share this excellent book with your little ones.

For Grade-Schoolers

Cobb, Vicki, *Supersuits* (Lippincott, 1975).
Surviving in extreme cold, underwater, in thin or nonexistent air, or fire, takes special clothing. A great "gee-whiz" text.

Inventions That Changed Our Lives (Walker).
This series (there are volumes on sewing machines, the telescope, dynamite, the elevator, the compass, and more) is a superb combination of biography, history, technology, and artful writing.

Lasky, Kathryn, *The Weaver's Gift* (Warne, 1980).
A homely story of how sheep are sheered and wool is woven into a blanket, by an excellent children's author.

Lauber, Patricia, *Too Much Garbage* (Garrard, 1974).
Despite the publication date, this volume tells just what kids need to know about problems of waste disposal and recycling.

Lewis, Alun, *Super Structures* (Viking, 1980).
An excellent civil engineering book for kids which focuses on the construction of modern support techniques.

National Geographic Society, *How Things Are Made* (National Geographic Society, 1981).

Baseballs, Band-Aids, kaleidoscopes, marbles—they were all made and this book tells you how.

Rawson, Christopher, *How Machines Work* (Usborne/Hayes, 1976).

A nice accompaniment to a building toy with motors and gears.

Schulz, Charles M., *Charlie Brown's Fifth Super Book of Questions and Answers: About All Kinds of Things and How They Work!* (Random House, 1981).

Volume 5, the best in this series, provides clear and appealing answers to questions commonly asked.

For Middle-Schoolers

Grummer, Arnold E., *Paper by Kids* (Dillon, 1980).

Using readily available materials and easy-to-construct equipment, this book tells you how to make paper at home.

MacGregor, Anne and Scott, *Domes: A Project Book,* (Lothrop, Lee & Shepard, 1981).

Not only provides excellent directions, but also provides historical and explanatory information not found in any of the dome kits.

St. George, Judith, *The Brooklyn Bridge: They Said It Couldn't Be Built* (Putnam, 1982).

This award-winning title masterfully captures the adventure and trauma of building New York's famous bridge.

Salvadori, Mario, *Building: The Fight Against Gravity* (Atheneum, 1979).

This book gives a clear sense of engineers as problem-solvers working with a variety of materials. If you've ever wondered why a building doesn't sink, or why skyscrapers don't fall, this one's for you.

Weiss, Harvey, *How to Be an Inventor* (Crowell, 1980).

Weiss answers questions for creative and ambitious kids by describing a variety of inventions and ways to design and patent your own plans.

Zubrowski, Bernie, *Messing Around with Drinking Straw Construction* (Little, Brown, 1981).

Tells children in grades 4 to 7 how to build houses, bridges, and other structures which illustrate fundamental engineering principles.

———, *Messing Around with Water Pumps and Siphons* (Little, Brown, 1981).

Hands-on activities to explore suction, compression, and the like.

Note: If you give a construction book as a gift, include the recommended pipe cleaners, drinking straws, or tubing—it's much like including batteries with a toy that requires them to work.

Index

SMALL CAPS indicate brand or trade names.

COPYRIGHT ACKNOWLEDGMENTS

Grateful acknowledgment is made for permission to reprint:

"Code of Practice on Animals in Schools" from the September 1980 issue of *Science and Children.* Copyright © 1980 by the National Science Teachers Association. Reprinted by permission.

Judges' guidelines adapted from judging criteria used by the Ohio Academy of Science from *Science Day Guide* by Joanne Zinsner Mann. Reprinted by permission of the Ohio Academy of Science.

"Teaching Your Children the Joys of Woodworking" by Richard Starr from the December 1983 issue of *Family Handyman Magazine.* Reprinted by permission of *Family Handyman Magazine,* The Webb Company, 1999 Shepard Road, St. Paul, MN 55116, © copyright 1985. All rights reserved.

Excerpt from *The New Encyclopedia of Science.* © 1982 New Encyclopedia of Science. Reprinted by permission of Raintree Publishers, Inc.

"How to Make a Terrarium" from *Beastly Neighbors* by Mollie Rights, illustrated by Kim Solga. Copyright © 1981 by Yolla Bolly Press. Reprinted by permission of Little, Brown and Company.

Photograph of HUMAN SKELETON is reprinted by permission of Albion Import Export Company.

ABOUT THE AUTHORS

WENDY SAUL, who holds a Ph.D. in Curriculum and Instruction from the University of Wisconsin, Madison, teaches in the Department of Education at the University of Maryland Baltimore County. ALAN R. NEWMAN, her chief science advisor and husband, received his Ph.D. in Chemistry from the University of Pennsylvania. They are the parents of Matthew and Eliza Newman-Saul, two tough critics.